LIFELINES™ FOR THE GMAT®

MATH WORKBOOK

113 LIFELINES YOU NEED TO KNOW

A FULL GMAT MATH PREP COURSE — THE CLASSROOM EXPERIENCE IN A BOOK

RON WOLDOFF & ELLEYNE KASE

EDUCATION FOR ADVANCEMENT™

thinktankSCHOLASTIC™ NATIONAL TEST PREP™

PHOENIX ARIZONA, USA

LifeLines for™ the GMAT® Math Workbook

Published by
Thinktank Scholastic™
13516 West Whitewood Drive
Sun City West Arizona, 85375
www.thinktankscholastic.com

Thinktank Scholastic™ also publishes its books in a variety of electronic formats. Some content that appears in print may not be available in electronic books.

For general information on our other products and services, please contact Thinktank Scholastic at 623-444-6707 or visit www.thinktankscholastic.com.

ISBN 978-0-9802242-3-8

Manufactured in the United States of America

thinktankSCHOLASTIC™

AN IMPRINT OF
Max WISE *Publications*™

About The Authors

Ron Woldoff completed his dual MBA-MIS degree from Arizona State University and San Diego State University, where he studied the culmination of business and technology. After graduating, he spent several years consulting for several Fortune-500 and local corporations, including Honeywell, American Express, APS, and Cox Communications.

In 2008, taking a break from the corporate world, Ron helped some students prepare for the GMAT, the GRE, and the SAT. With these subjects and his teaching services in demand, National Test Prep was established that counted private students along with Northern Arizona University and the internationally acclaimed Thunderbird School of Global Management as clients.

The world of publishing made a path to his door with Wiley and Sons in New York inviting him to author the flagship product "GRE for Dummies," published in 2011.

Ron is the founder of National Test Prep™ and creator of the LifeLine™ study approach. He also developed all the course materials, curricula, PowerPoint lessons, and printed materials for his education center.

On this, he remarked, "I applied my skills as an analyst to the exams by looking for arrangements, types of questions, and common traps. The key to answering a question correctly was in how to approach the problem. Starting from there, I designed the curricula accordingly—to great success.

"I love what I do. Helping people succeed is a good vocation. Plus, it's fun being the only person in the room who has the answers."

Elleyne Kase started out in advertising in the San Francisco Bay Area, California. She has many years of design experience in a variety of venues and media. She finds book design one of the most rewarding.

Elleyne is the creator of the Visual Quick Notes series for financial professionals, and she found the chance to expand into graduate test prep an exciting opportunity.

"The process of designing began as a total immersion into the GMAT world. From beginning to end I was the GMAT student making my way through the content, skill by skill, task by task. I attended the classes, worked the problems, and participated in the group solution-on-the-board process with Ron and the other students.

"The added challenge for me—my own GMAT level puzzle—was establishing a primarily verbal-based curriculum into concrete written form, and from there, creating an easy to follow self-study graphic framework, that could be used in book or other media.

"Starting from transcripts of class lectures, the content was molded and the graphic structure took form. The sheer level of hierarchy was more than usually encountered in a project which, combined with the variety of concepts, was the first level of challenge. Adding to that, drills and quizzes on right hand pages, solutions on left ones, made for demanding criteria indeed. Each of us had our own must-have list; which weren't happening, sent us back to the drawing board.

"In time we made our way to what you have in hand today and is what Ron, the most demanding-chooser-of-curriculum materials on the planet, now uses in his classes.

"We are both proud and confident that this study tool can well serve anyone wishing to take the GMAT and achieve their goals, as his students can attest to."

Acknowledgements

Ron Woldoff

For friends, I would like to thank Elleyne for recognizing the value of my classes and encouraging me to package them into a product for many more to use.

I would also like to thank Ken Krueger, Lionel Hummel, and Jaime Abromovitz, who offered invaluable guidance and feedback as I was getting things started.

For family, I would like to thank my mom, Sarah, incidentally the only other person that I know of who teaches her own material at a university campus, for all of her continued support. And more than anyone else, I would like to thank my wife, Leisah, for her faith in me and the work that I do. Finally, I'd like to thank our three beautiful boys, Zachary, Jadon, and Adam, for giving my work a purpose.

Elleyne Kase

I want to thank Ron for giving me this opportunity to enter his world of graduate exam prep, and team with him to create a new educational series. He was willing to take that great leap of faith and enter into my world of crafting an idea, into a finished, real-world, product with all the accompanying challenges.

A special thanks goes to Katja Rigter for her professional advice, and gratitude to my three wonderful children, Seth, Shane and Mimi, and my mother who stood by me, believing in me, encouraging me throughout the process.

GMAT Math Workbook

"The reason for time is to make sure everything doesn't happen at once."

– Albert Einstein – THEORETICAL PHYSICIST, (1879–1955)

Table of Contents [Brief]

Table of Contents [Full]

The Process

The Strategy

What the Pages Look Like

Headings
Tell exactly where you are in the book at any time.

Topics
Define the question type.

LifeLines
Identify and explain the task you need to know.

Drills & Quizzes
Test and practice your skills.

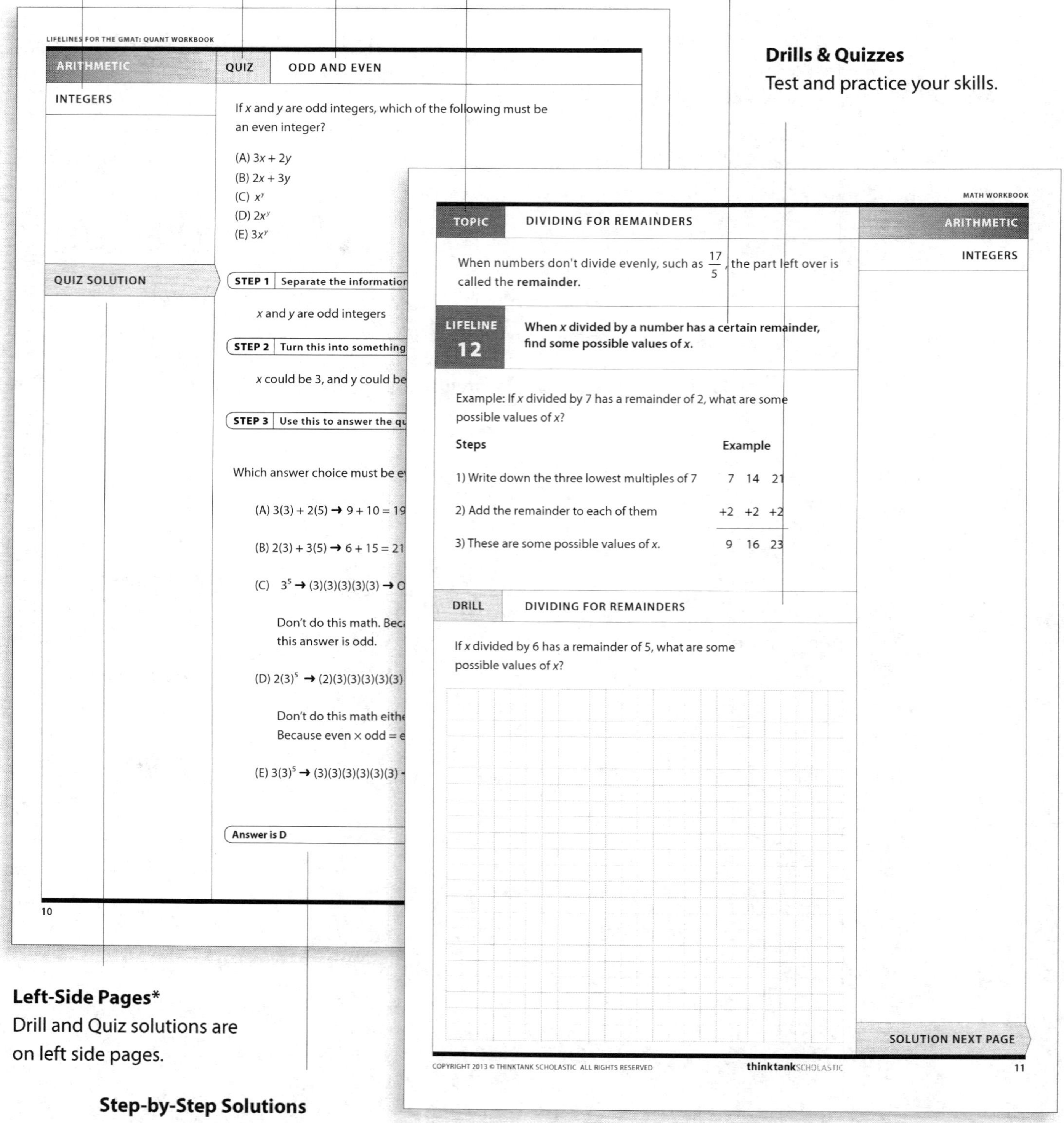
LIFELINES FOR THE GMAT: QUANT WORKBOOK

ARITHMETIC | QUIZ | ODD AND EVEN

INTEGERS

If x and y are odd integers, which of the following must be an even integer?

(A) $3x + 2y$
(B) $2x + 3y$
(C) x^y
(D) $2x^y$
(E) $3x^y$

QUIZ SOLUTION

STEP 1 Separate the information

x and y are odd integers

STEP 2 Turn this into something

x could be 3, and y could be

STEP 3 Use this to answer the qu

Which answer choice must be e

(A) 3(3) + 2(5) → 9 + 10 = 19

(B) 2(3) + 3(5) → 6 + 15 = 21

(C) 3^5 → (3)(3)(3)(3)(3) → O

Don't do this math. Bec
this answer is odd.

(D) $2(3)^5$ → (2)(3)(3)(3)(3)(3)

Don't do this math eith
Because even × odd = e

(E) $3(3)^5$ → (3)(3)(3)(3)(3)(3)

Answer is D

10

MATH WORKBOOK

TOPIC | DIVIDING FOR REMAINDERS | ARITHMETIC

INTEGERS

When numbers don't divide evenly, such as $\frac{17}{5}$, the part left over is called the **remainder**.

LIFELINE 12 — **When x divided by a number has a certain remainder, find some possible values of x.**

Example: If x divided by 7 has a remainder of 2, what are some possible values of x?

Steps	Example
1) Write down the three lowest multiples of 7	7 14 21
2) Add the remainder to each of them	+2 +2 +2
3) These are some possible values of x.	9 16 23

DRILL | DIVIDING FOR REMAINDERS

If x divided by 6 has a remainder of 5, what are some possible values of x?

SOLUTION NEXT PAGE

COPYRIGHT 2013 © THINKTANK SCHOLASTIC ALL RIGHTS RESERVED thinktankSCHOLASTIC 11

Left-Side Pages*
Drill and Quiz solutions are on left side pages.

Step-by-Step Solutions
Are broken down with clear illustrations and notations.

Right-Side Pages
Have LifeLines, Drills and Quizzes, with space to work the questions

* Some exceptions to this order of content do appear.

What the LifeLine Study Process Looks Like

Topics & LifeLines — Define & Explain Question Types

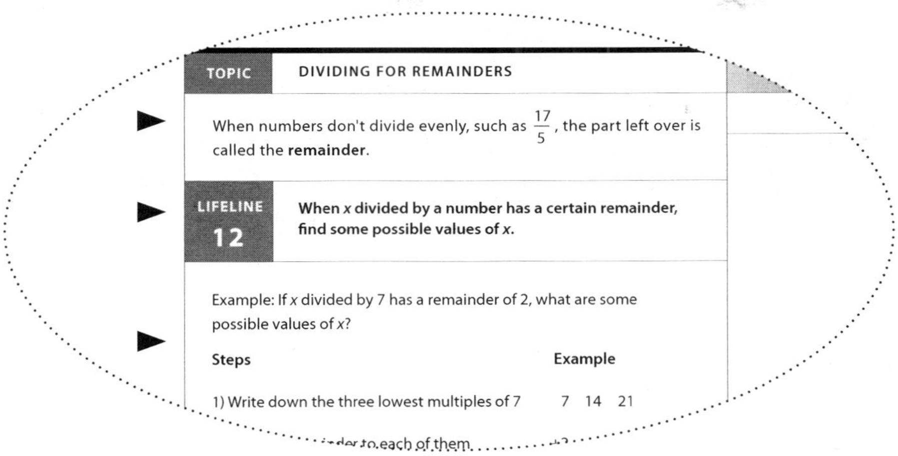
TOPIC | DIVIDING FOR REMAINDERS

When numbers don't divide evenly, such as $\frac{17}{5}$, the part left over is called the **remainder**.

LIFELINE 12 | **When x divided by a number has a certain remainder, find some possible values of x.**

Example: If x divided by 7 has a remainder of 2, what are some possible values of x?

Steps	Example
1) Write down the three lowest multiples of 7	7 14 21

Topics define the question type

LifeLines explain how to approach the question type or what to do when stuck.

Examples illustrate the LifeLine task.

Drills — Test and Build Basic Skills Based on Question Types

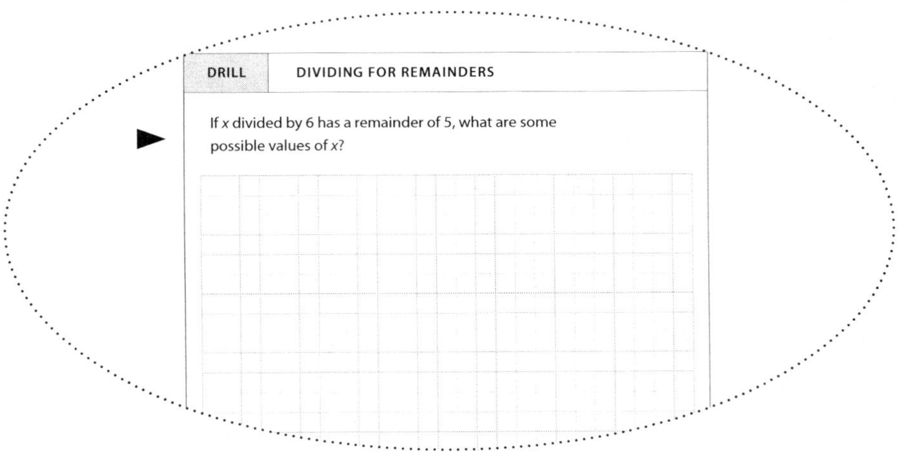
DRILL | DIVIDING FOR REMAINDERS

If x divided by 6 has a remainder of 5, what are some possible values of x?

Drills let you practice the basic skill set.

Drill Solutions give you a chance to check your work.

Quizzes — Expand and Build Advanced Skills Based on GMAT-Level Questions

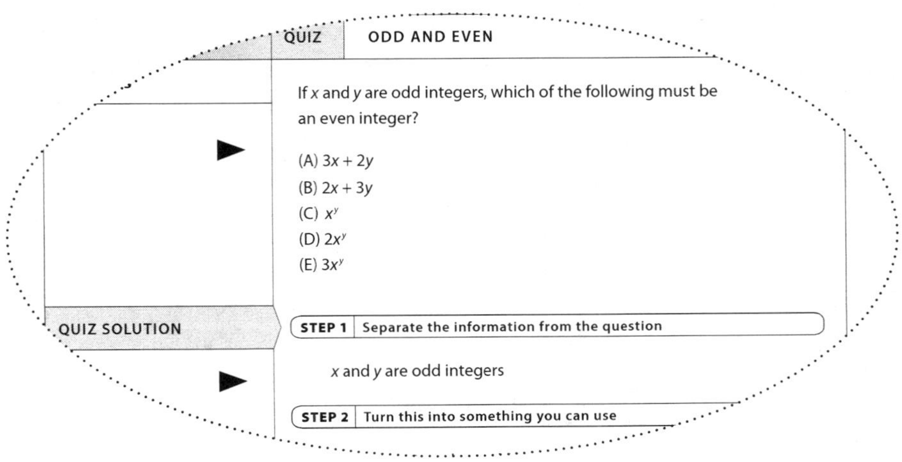
QUIZ | ODD AND EVEN

If x and y are odd integers, which of the following must be an even integer?

(A) $3x + 2y$
(B) $2x + 3y$
(C) x^y
(D) $2x^y$
(E) $3x^y$

QUIZ SOLUTION

STEP 1 | Separate the information from the question

x and y are odd integers

STEP 2 | Turn this into something you can use

GMAT-Level Quizzes show you what to expect with each question type.

Step-by-Step Solutions demonstrate the way to think through the question quickly and easily.

FYI

The Five Most Common GMAT Math Question Types

The GMAT features 37 math questions based on more than 200 math concepts. Below are examples of the five most commonly asked math question types on the GMAT.

1. Exponents

If $5^{11} \times 4^5 = 5 \times 10^k$, what is the value of k?

2. Functions

For all numbers y such that $y \neq 1$, if $f(y) = \frac{y^2+2}{y+1}$, then $\left(\frac{1}{f(y)}\right)\left(\frac{1}{f(2)}\right) =$

3. Work Problems

Four machines, each working at the same constant rate, together can complete a certain job in 12 days. How many machines, each working at the same constant rate, will be needed to complete the job in eight days?

4. Counting Methods

Two marbles are to be selected from a box. If there are 45 different possible selections of the two marbles, how many marbles are in the box?

5. Triangles

In the figure shown, if the measure of angle *ADB* is 20°, the measure of *CAD* is how many degrees less than the measure of *CBD*?

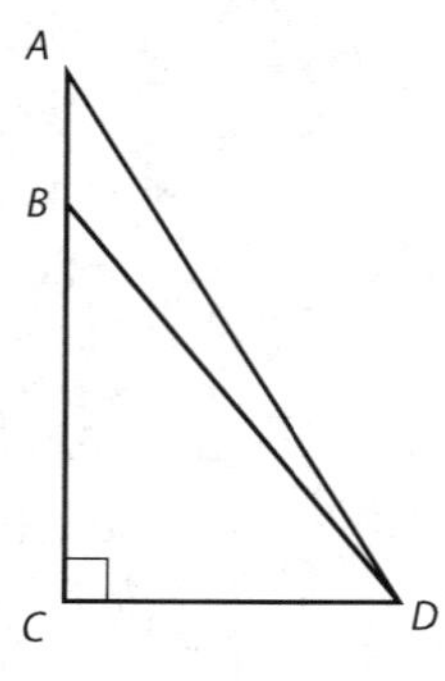

Your Time Line

Take a practice test now, to see your starting point.

The software is free from **mba.com**

GMAT score you HAVE

____

GMAT score you NEED

____

Date of your test

____

Study Math & Verbal	**Break**	**Study** Practice Tests and Review	**Test**
6 Weeks	**2 Weeks**	**3 Weeks**	**Final Week**

↑ Today ↑ Could repeat ←

LIFELINE 1 — **If you're not sure when to take your GMAT, check your timeline, and schedule the test now.**

You can always reschedule for a nominal fee, but **this way you know that you can take the GMAT at a time that works for you.**

LIFELINE 2 — **If you become exhausted, which is normal, take a two-week break from studying.**

The break is a vital part of your prep process. The GMAT is an intense, difficult test, and student burn-out is common. When you're exhausted, you don't perform as well as you could on the test. Students are always reluctant to take a break, but afterwards are glad they did.

After the break, jump back in for three weeks, and brush up on all your skills. It's normal to forget how to work a question when there are 200 questions, so this review period makes all the difference.

LIFELINE 3 — **When you think you're ready for the GMAT, take a practice test.**

Any mistake that you're going to make, such as not clicking to the end, should be made on the practice test and not the real thing. The benefit of the practice test comes from reviewing it afterwards. See the kinds of mistakes that you make and on what question types. Use the Wrong Question Tracker tear sheet at the back of this book, page 413.

Core strategy

How to study—

The first GMAT question answered.

Start on page 1, lesson 1. Use this workbook as your study guide to cover what's in the math section of the exam.

Don't worry—

If you forgot your math, you're in good company.

A lot of our students have forgotten their math, but it usually comes right back, and they do just fine. This book assumes that you're in that group. We refresh the basics (such as how to add fractions) and continue to the GMAT-level questions.

Four skills will get you through the GMAT

Question Skills

Test-Taking Skills

Stamina

Control

First—Question Skills

Knowing ... What the GMAT Will Ask and How To Solve at the GMAT Level

You will learn:

1. What the GMAT will ask ______________ in the Topics
2. How to work the question types ________ in the LifeLine and Examples
3. How to work GMAT-level questions _____ in the Drills and Quizzes

Doing well on the GMAT isn't about being smart; it's a skill, and like any other skill, it takes practice.

Second—Test-Taking Skills

These factors affect your score:

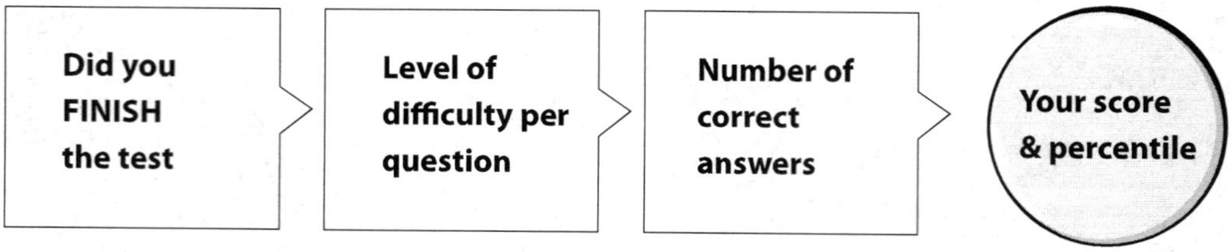

Test-taking skills are skills that you need to do well on the GMAT as a whole. These skills are just as important as knowing how to answer each question.

There are two test-taking skills in particular that are easy to learn and important to master: one is finishing the test, and the other is using the test's "adaptive" nature to your advantage.

Finish the test

LIFELINE 4	**When you have three minutes left, stop working and click to the end.**

Finishing the test means making sure you reach the last question before time runs out. You don't have to get the question right, but you do have to answer it. If you don't answer the last question, the GMAT significantly reduces your score.

What "finishing the test" looks like in practice

Below is an exact mock up of a review screen after some students took a practice test as a group. We took our time in the beginning, clarifying points and reviewing questions. On question 17—not even halfway—we had three minutes left, so we stopped working and started clicking. We made sure that we finished the section, even if it meant guessing on all those questions.

Question 1		Question 15		Question 28	Incorrect
Question 2		Question 16		Question 29	Incorrect
Question 3		Question 17	Incorrect	Question 30	Incorrect
Question 4		Question 18		Question 31	
Question 5		Question 19	Incorrect	Question 32	
Question 6		Question 20		Question 33	Incorrect
Question 7		Question 21	Incorrect	Question 34	Incorrect
Question 8		Question 22	Incorrect	Question 35	
Question 9		Question 23		Question 36	
Question 10		Question 24	Incorrect	Question 37	Incorrect
Question 11		Question 25	Incorrect		
Question 12		Question 26	Incorrect		
Question 13		Question 27	Incorrect		

We did this in both the math and the verbal sections: answered the first half correctly and clicked through the second half. We got a 700, which is a 90-percentile ranking.

Using the test's "adaptive" nature to your advantage.

The number of correct answers in the beginning of each section sets the level of difficulty and determines your score.

On question #1, The GMAT gives you a medium-level question. If you get it right, the test gives you a harder question. The difficulty level increases until you get a question wrong. Then the GMAT eases up until you get the questions right.

In this way, the GMAT "adapts" to your skill level, which is **why it's called an "adaptive" test.** By getting the early Math and Verbal questions right, you raise the bar and set the difficulty level high.

As an "adaptive" test, the level of difficulty is more important than the number of correct answers.

About halfway through the section, the GMAT has a good sense of your ability. Therefore, the earlier the question, the more it counts. These early questions are the same whether you're prepared or not. The difference is how challenging they are to you.

If you're prepared, these early questions will be the least challenging that you get. By the time the questions are too hard for you to answer correctly, you've set the bar high and achieved a high score. (Usually, at this point you're down to three minutes and are clicking to the end.)

If you're not prepared, the opposite is true. The early questions will be the most challenging that you get. Halfway through the section, when the GMAT finally makes the questions easy enough for you to answer correctly, you've set the bar low and earned a low score.

If you say you got only 12 of the 37 math questions wrong, what's important is whether they were at the beginning or the end of the section.

Rushing is the recipe for failure.

In the math section, you have 75 minutes to work 37 questions, which averages two minutes per question. However, pacing yourself at two minutes per question makes you more likely to rush through the math, make mistakes, and get the questions wrong.

LIFELINE 5	**Forget the two minute per question strategy. Take the time that you need to get the question right.**

Rushing and making mistakes aren't conducive to using the test's "adaptive" nature to your advantage. Instead, take the time that you need and get the question right.

By taking your time on the early questions, you're more likely to get them right. Then, when there are three minutes left and you've set the difficulty level high, you can stop working and click to the end. This is how you use the test's **"adaptive" nature to your advantage**.

If you want to spend time building speed for the exam, do so by working lots of math basics drills (multiplication, exponents, etc.) with a stopwatch. See the kinds of mistakes that you make under pressure, and then you can track and fix them.

You can still do well if you miss an early question.

Getting the first half right and clicking to the end is one way to get a good score, but it's not the only way. If you miss a few early questions, you can still recover and do well. If you're at question number 2, and you're stuck, don't let it kill you. It's OK to get a couple wrong in the beginning.

LIFELINE 6	**If you're stuck on a question, take a guess, move on, and recover later in the section.**

We've seen students get early questions wrong, recover later in the section, and score well. Getting the first half right and then clicking through is one of many ways to do well on the GMAT.

Also, the adaptive difficulty level of the test is NOT a method for you to assess your performance during the test. Getting two easy questions in a row doesn't mean you're doing poorly. It doesn't mean you're doing well. It doesn't mean anything. **Just answer the questions and keep going.**

The Math and Verbal sections are adaptive, but Integrated Reasoning is not.

Third—Stamina

Working these questions for four hours is not like working a job for four hours. This level of concentration is much higher than what you're used to.

How the time is allotted	
One Essay	30 minutes
12 Integrated Reasoning Questions	30 minutes
37 Math Questions	75 Minutes
41 Verbal Questions	75 Minutes

LIFELINE 7	**If you find yourself getting exhausted, build your stamina by studying in 3-hour stretches.**

The first 3-hour study session hurts, like a new workout, but you'll get used to it. By the third study session, you'll be fine.

LIFELINE 8	**On the morning of the exam, warm up with simple math and reading.**

Make sure the first question of the day is not a GMAT question. Use the **pre-test math warm-up worksheet** among the tear sheets in the back of this workbook, pages 409-412.

Fourth—Control

Test takers choke **all the time**. It's like being on stage for the first time. It doesn't matter how well you know the song or the speech: the first time you're up there—you're a mess. The second time is much better. Students who retook the test tell me that they were much more at ease the second time around.

LIFELINE 9	**Control your test-day anxiety by making the GMAT testing-center experience as familiar as possible.**

Think of this as a dress rehearsal. If you have stage fright, get up on that stage a few times, and you'll control your stage fright. On the GMAT, you can't stop your anxiety, but you can control it.

Here's how to make the testing experience more familiar:

1. Group up and compete with other test takers, so that you get used to the sounds of them working and the feeling of competition.

   Some testing centers provide earplugs or headphones. Find out exactly what your testing center provides and practice with these if this will help.

2. Leave all your personal items, including snacks and water, in another room.

3. Lowest score buys pizza.

   We've never actually made a student pay for everyone's pizza, but the idea introduces the competitive element that's missing when you take the practice test on your own.

   Students tell us that taking the practice test as a group, ordering in pizza, and reviewing the practice tests together is a great way to prepare for the real thing.

Here's an email from a student after taking the GMAT:

In regards to specific questions that we didn't cover I really didn't have any issue … One of the factors that I think makes it tough for students is simply the testing environment as it is intimidating with the palm scans and hand raises any time you wish to leave the workstation. It makes an already intense situation that much more intense …

Don't be surprised when your personal items, including your cell phone, wallet/purse, wristwatch, and water are placed out of reach, usually in a locker. If you're not expecting this level of security in the testing center, it can make you uncomfortable.

Also, if you're not used to taking the test without sipping on water, this can be a distraction. Prepare for this by practicing without your water or cell phone nearby.

5 Math Lessons

Start Here

The lessons are progressive, becoming more challenging as you go.

Arithmetic

Arithmetic

Lesson 1

Setting the foundation for the

concepts in all other lessons.

TOPIC	FACTORS AND MULTIPLES	ARITHMETIC
		INTEGERS

An **integer** is any whole number, positive or negative, including 0.

A **factor** is an integer from dividing.

For example, factors of 21 are {1, 3, 7, 21}

Every number is a **factor** of itself.

A **multiple** is an integer from multiplying.

For example, multiples of 6 are {6, 12, 18 ...}

Every number is a **multiple** of itself.

LIFELINE 10

When *x* is a multiple of a number, use the three lowest multiples of the number.

Given	**Use**
x is a multiple of 3	3, 6, or 9 for *x*

QUIZ 1 FACTORS AND MULTIPLES

If the positive integer *x* is a multiple of 4, and the positive integer *y* is a multiple of 9, then the positive integer *xy* must be a multiple of which of the following?

I. 6
II. 8
III. 27

(A) I only
(B) I and II only
(C) I and III only
(D) II and III only
(E) I, II, and III

SOLUTION NEXT PAGE

ARITHMETIC | **QUIZ 1** | **FACTORS AND MULTIPLES**

INTEGERS

If the positive integer x is a multiple of 4, and the positive integer y is a multiple of 9, then the positive integer xy must be a multiple of which of the following?

I. 6
II. 8
III. 27

(A) I only
(B) I and II only
(C) I and III only
(D) II and III only
(E) I, II, and III

SOLUTION

STEP 1 | **Separate the information from the question**

x is a multiple of 4, and y is a multiple of 9

STEP 2 | **Turn this into something you can use**

If x is a multiple of 4, then x could be 4, 8, or 12

If y is a multiple of 9, then y could be 9, 18, or 27

STEP 3 | **Use this to answer the question**

xy (x times y) could be 36, 72, or 108

Check the statements one at a time.

6 yes: all these numbers are multiples of 6

8 no: some of these numbers are multiples of 8, but not all of them are.

27 no: not all the numbers are multiples of 27

Answer is A

Sometimes the question asks which number COULD be or COULD NOT be a multiple of xy.

This question asks which MUST be.

A certain integer n is a multiple of both 3 and 4. Which of the following must be true?

I. n is an even integer
II. n is equal to 12
III. n is a multiple of 6

(A) III only
(B) I and II only
(C) I and III only
(D) II and III only
(E) I, II, and III

SOLUTION NEXT PAGE

ARITHMETIC | **QUIZ 2** | **FACTORS AND MULTIPLES**

INTEGERS

A certain integer n is a multiple of both 3 and 4. Which of the following must be true?

I. n is an even integer
II. n is equal to 12
III. n is a multiple of 6

(A) III only
(B) I and II only
(C) I and III only
(D) II and III only
(E) I, II, and III

SOLUTION

STEP 1 **Separate the information from the question**

A certain integer n is a multiple of both 3 and 4.

STEP 2 **Turn this into something you can use**

n could be 12, 24, or 36.

STEP 3 **Use this to answer the question**

Knowing n could be 12, 24, or 36, check the statements one at a time.

I. n is an even integer: yes
II. n is equal to 12: no
III. n is a multiple of 6: yes

Answer is C

TOPIC	ODD AND EVEN	ARITHMETIC
		INTEGERS

The GMAT turns a simple concept, such as odd times even, into a challenging question.

LIFELINE 11

When x is odd or even, use real odd or even numbers, such as 2 or 3.

For Example	Use
(odd)(odd) = odd	$3 \times 3 = 9$
(odd)(even) = even	$3 \times 2 = 6$
(even)(even) = even	$2 \times 2 = 4$

QUIZ — ODD AND EVEN

If x and y are odd integers, which of the following must be an even integer?

(A) $3x + 2y$

(B) $2x + 3y$

(C) x^y

(D) $2x^y$

(E) $3x^y$

SOLUTION NEXT PAGE

ARITHMETIC | QUIZ | ODD AND EVEN

INTEGERS

If x and y are odd integers, which of the following must be an even integer?

(A) $3x + 2y$

(B) $2x + 3y$

(C) x^y

(D) $2x^y$

(E) $3x^y$

SOLUTION

STEP 1 Separate the information from the question

x and y are odd integers

STEP 2 Turn this into something you can use

x could be 3, and y could be 5

STEP 3 Use this to answer the question

Which answer choice must be even?

(A) 3(3) + 2(5) → 9 + 10 = 19 → Odd

(B) 2(3) + 3(5) → 6 + 15 = 21 → Odd

(C) 3^5 → (3)(3)(3)(3)(3) → Odd

Don't do this math. Because odd × odd = odd, this answer is odd.

(D) $2(3)^5$ → (2)(3)(3)(3)(3)(3) → Even

Don't do this math either.
Because even × odd = even, this answer is even.

(E) $3(3)^5$ → (3)(3)(3)(3)(3)(3) → Odd

Answer is D

TOPIC	DIVIDING FOR REMAINDERS	ARITHMETIC
		INTEGERS

When numbers don't divide evenly, such as $\frac{17}{5}$, the part left over is called the **remainder**.

LIFELINE 12

When *x* divided by a number has a certain remainder, find some possible values of *x*.

Example: If *x* divided by 7 has a remainder of 2, what are some possible values of *x*?

Steps	Example		
1) Write down the three lowest multiples of 7	7	14	21
2) Add the remainder to each of them	+2	+2	+2
3) These are some possible values of *x*.	9	16	23

DRILL — DIVIDING FOR REMAINDERS

If *x* divided by 6 has a remainder of 5, what are some possible values of *x*?

SOLUTION NEXT PAGE

ARITHMETIC

INTEGERS

SOLUTION

DRILL — DIVIDING FOR REMAINDERS

If x divided by 6 has a remainder of 5, what are some possible values of x?

Steps	Example
1) Write down the three lowest multiples of 6	6 12 18
2) Add the remainder to each of them	+5 +5 +5
3) These are some possible values of x.	11 17 23

DETAIL — DIVIDING FOR REMAINDERS

$$\frac{25}{4} = 6 \text{ remainder } 1$$

$$= 6\frac{1}{4}$$

$$= 6.25$$

The decimal .25 comes from the remainder 1 placed over the divisor 4

$$.25 = \frac{1}{4} \begin{array}{l} \leftarrow \text{remainder} \\ \leftarrow \text{divisor} \end{array}$$

QUIZ 1 **DIVIDING FOR REMAINDERS**

ARITHMETIC

INTEGERS

If x divided by 3 has a remainder of 1, and x divided by 2 has a remainder of 1, then which of the following could be the value of x?

(A) 4
(B) 5
(C) 6
(D) 7
(E) 8

SOLUTION NEXT PAGE

ARITHMETIC

INTEGERS

QUIZ 1

DIVIDING FOR REMAINDERS

If x divided by 3 has a remainder of 1, and x divided by 2 has a remainder of 1, then which of the following could be the value of x?

(A) 4
(B) 5
(C) 6
(D) 7
(E) 8

SOLUTION

STEP 1 Separate the information from the question

x divided by 3 has a remainder of 1

x divided by 2 has a remainder of 1

STEP 2 Turn this into something you can use

If x divided by 3 has a remainder of 1,

x could be 4, 7, 10, ...

If x divided by 2 has a remainder of 1

x could be 1, 3, 5, 7, ...

STEP 3 Use this to answer the question

The only number in both lists is 7

Answer is D

QUIZ 2 DIVIDING FOR REMAINDERS

If the positive integer n is a multiple of 5 and has a remainder of 2 when divided by 9, what is one possible value of n?

(A) 10
(B) 11
(C) 15
(D) 18
(E) 20

SOLUTION NEXT PAGE

ARITHMETIC | **QUIZ 2** | **DIVIDING FOR REMAINDERS**

INTEGERS

If the positive integer *n* is a multiple of 5 and has a remainder of 2 when divided by 9, what is one possible value of *n*?

(A) 10
(B) 11
(C) 15
(D) 18
(E) 20

SOLUTION

STEP 1 Separate the information from the question

n is a multiple of 5 and has a remainder of 2 when divided by 9.

STEP 2 Turn this into something you can use

If *n* is a multiple of 5

it could be 5, 10, 15, 20, or 25.

If it has a remainder of 2 when divided by 9

it could be 11, 20, or 29.

STEP 3 Use this to answer the question

The only number in both lists is 20

Answer is E

TOPIC | **PRIME NUMBERS AND FACTORING** | **ARITHMETIC**

INTEGERS

A **Prime Number** has three rules:

1) It is whole
2) It is positive
3) It has exactly two factors

2 is the only even prime number
1 is not prime, as it has only one factor
0 is not prime, as it's neither positive nor has two factors

You won't have to discern prime numbers above 20.
Knowing the first few prime numbers is fine: 2, 3, 5, 7, 11, 13, 17, and 19.

> Some math books consider 1 to be prime. However, the GMAT does not, and it expects you to know that.

All positive integers (other than 1) are either prime or made up of a single, unique set of prime numbers. For example:

$12 = 2 \times 2 \times 3$
$24 = 2 \times 2 \times 2 \times 3$
$36 = 2 \times 2 \times 3 \times 3$

LIFELINE 13

When reducing a fraction, don't use long division: instead, factor and cancel.

To reduce $\frac{144}{16}$, long division works $16\overline{)144}$ but not well.

Instead, leave it as a fraction, and **factor** the numbers out so they cancel.

1. Cancel each 12 over 4:

```
12  12           3    3
                12̸   12̸
  \ /              \ /
  144̸              144̸
  ---      =       ---
  16̸               16̸
  / \              / \
 4   4            4̸   4̸
```

2. Multiply the remaining 3's: $3 \times 3 = 9$

> Even though $16 = 8 \times 2$, 4×4 works better for cancelling with 12 and 12.
>
> You don't have to factor to the primes—just factor to numbers that cancel.

ARITHMETIC

INTEGERS

LIFELINE 14

When factoring a number with zeros, such as 300, always start with the zeros: 3 × 100

To reduce $\frac{1200}{48}$, avoid long division $48\overline{)1200}$

Instead, leave it as a fraction, and **factor** the numbers out so they cancel.

> A fraction becomes an integer when all the numbers on the bottom cancel out.

The 12's cancel, and the remaining fraction is easy to solve.

$$\frac{1200}{48} = \frac{\cancel{12} \times 100}{\cancel{12} \times 4} = \frac{100}{4} = 25$$

DRILL — REDUCING FRACTIONS BY FACTORING

Reduce $\frac{46{,}000}{115}$ by factoring.

SOLUTION NEXT PAGE

DRILL — REDUCING FRACTIONS BY FACTORING

ARITHMETIC

INTEGERS

DRILL SOLUTION

Reduce $\frac{46,000}{115}$ by factoring.

When factoring numbers with zeros, factor the zeros first.

$$2 \times 200 = 400$$

$$\frac{46,000 \;(\cancel{46} \cdot \cancel{1000})}{115 \;(\cancel{23} \cdot \cancel{5})}$$

Because 115 ends in 5, you know that it's divisible by 5.

$$5\,\overline{)115}\; = 23$$

It's OK to use long division sparingly.

QUIZ 1 — REDUCING FRACTIONS BY FACTORING

If the product of the integers n, p, q, and r is 210, and $1 < n < p < q < r$, what is the value of $n + p$?

(A) 2
(B) 3
(C) 5
(D) 7
(E) 21

SOLUTION NEXT PAGE

ARITHMETIC | **QUIZ 1** | **REDUCING FRACTIONS BY FACTORING**

INTEGERS

If the product of the integers n, p, q, and r is 210, and $1 < n < p < q < r$, what is the value of $n + p$?

(A) 2
(B) 3
(C) 5
(D) 7
(E) 21

SOLUTION

STEP 1 Separate the information from the question

The product of the integers n, p, q, and r is 210, and $1 < n < p < q < r$

STEP 2 Turn this into something you can use

Factor the 210 to find n, p, q, and r

```
    210
    / \
  21   10
  /\   / \
 3  7 2   5
```

Because $1 < n < p < q < r$,

$n = 2, \ p = 3, \ q = 5, \text{ and } r = 7$

STEP 3 Use this to answer the question

$n + p \rightarrow 2 + 3 = 5$

Answer is C

If y is a positive integer less than 25, and $\frac{10y}{12}$ is an integer, then y has how many different prime factors?

(A) One
(B) Two
(C) Four
(D) Five
(E) Seven

SOLUTION NEXT PAGE

ARITHMETIC | **QUIZ 2** | **REDUCING FRACTIONS BY FACTORING**

INTEGERS

If y is a positive integer less than 25, and $\frac{10y}{12}$ is an integer, then y has how many different prime factors?

SOLUTION

STEP 1 Separate the information from the question

y is less than 25, and $\frac{10y}{12}$ is an integer

STEP 2 Turn this into something you can use

> **Remember**
> A fraction becomes an integer when the numbers on the bottom cancel out, which in this case are 2 and 3.

Reduce the fraction $\frac{10y}{12} = \frac{5y}{6}$

Factor out the 6 → 2 3

This means that y has to have 2 and 3 as factors to cancel.
It could have other factors, but it at least needs a 2 and a 3.

> **Could y have a 5?**
> No, because $2 \times 3 \times 5 = 30$, which is greater than 25.

***y* could equal:**
6 (2×3), 12 $(2 \times 2 \times 3)$, 18 $(2 \times 3 \times 3)$, or 24 $(2 \times 2 \times 3 \times 3)$.

No matter what: the 2 and 3 on the bottom cancel, and the numbers that remain on top leave an integer.

STEP 3 Use this to answer the question

With any of these possible values for y,
it still has two different prime factors: 2 and 3.

Answer is B

TOPIC DIVISIBILITY

ARITHMETIC

INTEGERS

Factoring is the key to answering some of the trickier questions, as we'll see later in this book. Here are some easy ways to tell what a large number is divisible by.

LIFELINE 15

When dividing a number, use the number properties instead of long division.

DETAIL **Checking divisibility by 2, 4 or 8 — look at the last digits**

2 — look at the **last digit** in the number.

If last digit is divisible by 2 [13,57**4**] Or it's a zero [31,54**0**]

The number is divisible by 2, regardless of the other numbers.

If the number ends with 0, you can factor out a 10, which is divisible by 2.

4 — look at the **last two digits** in the number.

If last 2 digits are divisible by 4 [13,5**32**] Or they're zeros [31,7**00**]

The number is divisible by 4, regardless of the other numbers.

If the number ends with 00, you can factor out a 100, which is divisible by 4.

8 — look at the **last three digits** in the number.

If last 3 digits are divisible by 8 [17,**104**] Or they're zeros [33,**000**]

The number is divisible by 8, regardless of the other numbers.

If the number ends with 000, you can factor out a 1000, which is divisible by 8.

DETAIL **Checking divisibility by 3 or 9 — add up the digits**

3 — If digits added are divisible by 3, [61,755 ➜ 6 + 1 + 7 + 5 + 5 = 24]

Because 24 is divisible by 3

The entire number is divisible by 3

9 — If digits are divisible by 9, [374,625 ➜ 3 + 7 + 4 + 6 + 2 + 5 ➜ 27]

Because 27 is divisible by 9

The entire number is divisible by 9

DETAIL **Checking divisibility by 5 or 10 — look at the last digit**

5 and 10 — If the number ends with 5 or 0, then the number is divisible by 5 or 10

ARITHMETIC

INTEGERS

DETAIL **Checking divisibility by 6 or 12 — factor the 6 or 12**

6 — Check 2 and 3.

If last digit is divisible by 2 [5,17**2** ➜ 2 ÷ 2 = **1**]

And if sum of digits is divisible by 3 [5,172 ➜ 5 + 1 + 7 + 2 = **15**]

The number is divisible by 6.

> Knowing this, you could check divisibility by much larger numbers.
>
> For example, an integer divisible by both 9 and 5 is divisible by 45, such as 2,610.

12 — Check 4 and 3

If last 2 digits are divisible by 4 [18,732 ➜ 32 ÷ 4 = **8**]

And if sum of digits is divisible by 3 [18,732 ➜ 1 + 8 +7 + 3 + 2 = **21**]

The number is divisible by 12.

DETAIL **Checking divisibility by 7**

7 — No shortcut method

DETAIL **Checking divisibility by 11— add up every other digit**

11 —

Add up every other digit: [181,665 ➜ 1 + 1 + 6 = 8]

Then add up the remaining digits: [181,665 ➜ 8 + 6 + 5 = 19]

Then subtract the sums: [8 – 19 = –11]

If you get 11, –11, or a multiple of 11 (such as 22), or zero ...

The number is divisible by 11.

> In one question, the number 143 comes up, and students think it's prime.
>
> Actually, it's divisible by 11.

TOPIC | **MULTIPLYING FOR THE "UNITS DIGIT"** | **ARITHMETIC**

INTEGERS

When you multiply two integers, the "units" digits of the answer comes directly from the "units" digits of the numbers multiplied.

For the number 1 , 2 3 **4** . 5 6 , the "units digit" is **4**.

1 = thousands digit (one thousand)
2 = hundreds digit (two hundred)
3 = tens digit (thirty)
4 = units digit (four)
5 = tenth digit (five tenths)
6 = hundredth digit (six hundredths)

1 , 2 3 **4** . 5 6

LIFELINE 16

When finding the "units" digit of the answer, use the "units" digits of the numbers you multiplied.

Steps	**Example**
When multiplying two numbers for the units digit	123,456 × 789,012
1. Cross off the non-units digits	~~123,45~~6 × ~~789,01~~2
2. Multiply the numbers that remain	6 × 2 = 12
3. Cross off the new non-units digit	~~1~~2
And you have the answer	2

DRILL | **MULTIPLYING FOR THE "UNITS DIGIT"**

1) Find the units digit of 27^3

2) Find the units digit of $(49^4)(37^2)(23^3)$

SOLUTION NEXT PAGE

ARITHMETIC | **DRILL** | **MULTIPLYING FOR THE "UNITS" DIGIT**

INTEGERS

SOLUTION

Find the units digit of 27^3

Steps	Example
When multiplying numbers for the units digit	$27 \times 27 \times 27$
1. Cross off the non-units digits	$\not{2}7 \times \not{2}7 \times \not{2}7$
2. Multiply the numbers that remain	$7 \times 7 = \not{4}9$; $9 \times 7 = 63$
3. Cross off the new non-units digit	$\not{6}3$
And you have the answer	3

Find the units digit of $(49^4)(37^2)(23^3)$

Steps	Example
When multiplying numbers for the units digit	$(49^4)(37^2)(23^3)$
Cross off the non-units digits	$(\not{4}9^4)(\not{3}7^2)(\not{2}3^3)$
Multiply the numbers that remain	$(9^4)(7^2)(3^3)$
The units digit of 9^4, 7^2, and 3^3 are 1, 9, and 7, respectively.	$1 \times 9 \times 7 = \not{6}3 \rightarrow 3$
And you have the answer	3

What is the units digit of 7^{12} ?

(A) 1
(B) 3
(C) 5
(D) 7
(E) 9

SOLUTION NEXT PAGE

ARITHMETIC

INTEGERS

QUIZ

MULTIPLYING FOR THE "UNITS DIGIT"

What is the units digit of 7^{12} ?

(A) 1
(B) 3
(C) 5
(D) 7
(E) 9

SOLUTION

STEP 1 Separate the information from the question

7^{12}

STEP 2 Turn this into something you can use

$$7^{12} = 7 \times 7 \times 7 \times 7 \times 7 \times 7 \times 7 \times 7 \times 7 \times 7 \times 7 \times 7$$

STEP 3 Use this to answer the question

You could multiply all the 7's together, dropping the tens digits along the way, and get the correct answer. However, that's a lot of math, so instead look for the pattern.

$$7 \times 7 \times 7 \times 7 \times 7 \times 7 \times 7 \times 7 \times 7 \times 7 \times 7 \times 7$$

$$= 49 \times 49 \times 49 \times 49 \times 49 \times 49$$

Drop the tens digits $= 9 \times 9 \times 9 \times 9 \times 9 \times 9$

$$= 81 \times 81 \times 81$$

Drop the tens digits again $= 1 \times 1 \times 1$

$$= 1$$

Answer is A

TOPIC	CONSECUTIVE INTEGERS	ARITHMETIC

There are three common types of consecutive integers.

Standard: {... –2, –1, 0, 1, 2, 3, ...}
Even: {... –4, –2, 0, 2, 4, 6, ...}
Odd: {... –3, –1, 1, 3, 5, 7, ...}

Most questions are based on **even** consecutive integers, but these lifelines work for **all** consecutive integers.

> The most common mistake for this question type is missing the word "even."

LIFELINE 17 — When working with consecutive integers, find the average.

Fortunately, the average is easy to find. The way that you find the average depends on what the question tells you.

Steps	**Example**
If you have all the numbers: The average is the middle number.	{22, 24, 26, 28, <u>30</u>, 32, 34, 36, 38} 30 is the average
Or the average of the two middle numbers.	{13, 15, 17, 19} ∧ 16 is the average

> To find the average of 15 and 17, don't add and divide by 2. Instead, use the number that goes in between: 16

If you have first and last numbers:	All consecutive odd integers from 15 to 35
Average the first and last numbers	$\frac{15+35}{2}$ = 25 is the average

If you have sum of the integers:	The sum of 10 consecutive even integers is 450
Divide the sum by the number of integers	$\frac{450}{10}$ = 45 is the average

36 38 40 42 44 | 46 48 50 52 54 = 450
45

ARITHMETIC

INTEGERS

TOPIC: FINDING ALL THE CONSECUTIVE INTEGERS

The number of integers, the average, and the type (standard, odd, or even) are all that you need to find all the consecutive integers.

LIFELINE 18

When finding all the consecutive integers, start with a dash for each integer.

Example: If 7 consecutive odd integers have an average of 19, what are all the integers?

Steps	**Example**
1. Draw 7 dashes for the 7 integers	_ _ _ _ _ _ _
2. Write in the average	_ _ _ 19 _ _ _
3. Knowing the consecutive integers are odd, write in the other integers	13 15 17 19 21 23 25

DRILL: FINDING ALL THE CONSECUTIVE INTEGERS

8 consecutive even integers have an average of 25. What are all the integers?

SOLUTION NEXT PAGE

DRILL	FINDING ALL THE CONSECUTIVE INTEGERS

8 consecutive even integers have an average of 25.
What are all the integers?

Steps	Example
1. Draw 8 dashes for the 8 integers	_ _ _ _ _ _ _ _
2. Write in the average	_ _ _ _ V _ _ _ _ 25
3. Knowing the consecutive integers are even, write in the integers above and below the average	_ _ _ 24 V 26 _ _ _ 25
4. Write in the remaining integers	18 20 22 24 V 26 28 30 32 25

Because 25 isn't one of the integers, it goes below the dashes.

QUIZ 1	FINDING ALL THE CONSECUTIVE INTEGERS

If the sum of five consecutive even integers is 100, the fourth integer is

(A) 21
(B) 22
(C) 23
(D) 24
(E) 25

SOLUTION NEXT PAGE

ARITHMETIC | QUIZ 1 | FINDING ALL THE CONSECUTIVE INTEGERS

INTEGERS

If the sum of five consecutive even integers is 100, the fourth integer is

(A) 21
(B) 22
(C) 23
(D) 24
(E) 25

SOLUTION

STEP 1 Separate the information from the question

The sum of five consecutive even integers is 100.

STEP 2 Turn this into something you can use

1. Draw 5 dashes for the 5 integers — __ __ __ __ __

2. Find the average — $\frac{100}{5} = 20$

3. Write in the average — __ __ 20 __ __

4. Knowing the consecutive integers are even, write in the remaining integers — 16 18 20 22 24

STEP 3 Use this to answer the question

The fourth integer is 22

Answer is B

The average (arithmetic mean) of eight consecutive even integers is 51. What's the highest of these integers?

(A) 54
(B) 55
(C) 56
(D) 57
(E) 58

SOLUTION NEXT PAGE

ARITHMETIC

INTEGERS

QUIZ 2 FINDING ALL THE CONSECUTIVE INTEGERS

The average (arithmetic mean) of eight consecutive even integers is 51. What's the highest of these integers?

(A) 54
(B) 55
(C) 56
(D) 57
(E) 58

SOLUTION

STEP 1 Separate the information from the question

The average of eight consecutive even integers is 51.

STEP 2 Turn this into something you can use

1. Draw 8 dashes for the 8 integers — _ _ _ _ _ _ _ _

2. Write in the average — _ _ _ _ V _ _ _ _ (51)

3. Knowing the consecutive integers are even, write in the integers above and below the average — _ _ _ 50 V 52 _ _ _ (51)

4. Write in the remaining integers — 42 44 46 50 V 52 54 56 58 (51)

STEP 3 Use this to answer the question

The highest integer is 58.

Answer is E

TOPIC	FINDING THE SUM OF CONSECUTIVE INTEGERS	ARITHMETIC
		INTEGERS

When the GMAT asks for the sum of consecutive integers, don't actually add them up. There's an easier way.

LIFELINE 19

To find the sum of consecutive integers, multiply the average by the number of terms.

To find the sum of 22, 24, 26, 28, 30, 32, 34, 36, 38

1. Find the average — 30
2. Multiply by the number of terms — × 9

$$30 \times 9 = 270$$

To find the sum of all consecutive odd integers from 15 to 35

1. Find the average — 25
2. Multiply by the number of terms — × 11

$$25 \times 11 = 275$$

LIFELINE 20

To quickly count the consecutive integers, start by subtracting the lowest from the highest.

When counting consecutive even or odd integers

1. Subtract the lowest from the highest — $35 - 15 = 20$
2. Divide this number in half — $\frac{20}{2} = 10$
3. Add 1 — $10 + 1 = 11$

Though this lifeline works, you can also just count the integers.

When counting consecutive standard integers

1. Subtract the lowest from the highest — $100 - 50 = 50$
2. Add 1 — $50 + 1 = 51$

DRILL	FINDING THE SUM OF CONSECUTIVE INTEGERS

1) What is the sum of the consecutive even integers from 90 to 110?

2) What is the sum of the consecutive even integers from 892 to 908?

SOLUTION NEXT PAGE

ARITHMETIC

INTEGERS

SOLUTION

DRILL — FINDING THE SUM OF CONSECUTIVE INTEGERS

1) What is the sum of the consecutive even integers from 90 to 110?

Steps	Example
1. Find the average	100
2. Multiply by the number of terms	× 11
	1,100

2) What is the sum of the consecutive even integers from 892 to 908?

Steps	Example
1. Find the average	900
2. Multiply by the number of terms	× 9
	8,100

DETAIL — FINDING THE SUM OF CONSECUTIVE INTEGERS

You might recall this formula for finding the sum of consecutive integers:

$$\frac{n(n+1)}{2}$$

Where n represents the number of integers.
This formula only works if

1) The consecutive integers are standard, not even or odd

However, most consecutive integer questions on the GMAT are based on even, not standard, consecutive integers.

2) The first integer is 1

However, the consecutive integers usually begin higher than 1.

For these reasons, the formula doesn't help you on the GMAT, and we recommend using the Lifeline instead.

TOPIC	FRACTIONS	ARITHMETIC
		NON-INTEGERS

Though you probably were fine with fractions back in high school, most students haven't touched them since. Here's a refresher.

LIFELINE 21

When working with fractions, even just adding or subtracting, give the fractions a common denominator.

Example: $$\frac{1}{5} + \frac{1}{9} =$$

1. Find the common denominator $$\frac{1(9)}{5(9)} + \frac{1(5)}{5(5)} \rightarrow \frac{9}{45} + \frac{5}{45}$$

2. Add the numerators $$\frac{9+5}{45} = \frac{14}{45}$$

LIFELINE 22

When working with fractions, cancel what you can.

Example — multiplying: $$\frac{2}{5} \times \frac{3}{4} =$$

1. Cancel out what you can $$\frac{\cancel{2}}{5} \times \frac{3}{\cancel{4}} \rightarrow \frac{1}{5} \times \frac{3}{2}$$

2. Multiply straight across $$\frac{1}{5} \times \frac{3}{2} \rightarrow \frac{1 \times 3}{5 \times 2} \rightarrow \frac{3}{10}$$

LIFELINE 23

When dividing fractions, flip the bottom fraction and multiply.

Example: $$\frac{\frac{2}{3}}{\frac{5}{7}}$$

1. Flip the bottom fraction

$$\frac{2}{3} \times \frac{7}{5} =$$

2. Multiply $$\frac{2}{3} \times \frac{7}{5} \rightarrow \frac{2 \times 7}{3 \times 5} = \frac{14}{15}$$

ARITHMETIC

NON-INTEGERS

LIFELINE 24

When two fractions equal each other, and one has an *x*, cross-multiply.

To cross multiply two fractions, multiply the top of one fraction by the bottom of the other.

$$\frac{2}{3} = \frac{4}{3x}$$

$$\frac{2}{3} \times \frac{4}{3x}$$

$$6x = 12$$

$$x = 2$$

Cross multiplying also works, but this is faster.

When *x* is on the top of one of the fractions, give the fractions a common denominator.

$$\frac{2}{3} = \frac{x}{6}$$

$$\frac{2(2)}{3(2)} = \frac{x}{6}$$

$$\frac{4}{6} = \frac{x}{6}$$

$$x = 4$$

When there's only one fraction, multiply by the denominator.

$$\frac{x}{3} = 12$$

$$\frac{\cancel{(3)}x}{\cancel{3}} = 12(3)$$

$$x = 36$$

DRILL | FRACTIONS | ARITHMETIC

NON-INTEGERS

Solve for x

1. $\frac{2}{30} = \frac{1}{5x}$

$10x = 30$

$\frac{10x}{10} = \frac{30}{10}$

$x = 3$

2. $\frac{3}{10} = \frac{x}{50}$

$10x = 150$

$\frac{10x}{10} = \frac{150}{10}$

$x = 15$

3. $\frac{2x}{5} = \frac{10}{1}$

$2x = 50$

$\frac{2x}{2} = \frac{50}{2}$

$x = 25$

ARITHMETIC | DRILL | FRACTIONS

NON-INTEGERS

SOLUTION

Steps	Example
1. When x is on the bottom, cross multiply.	$\frac{2}{30} = \frac{1}{5x}$ $10x = 30$ $x = 3$
2. When x is on the top, use common denominators.	$\frac{3}{10} = \frac{x}{50}$ $\frac{15}{50} = \frac{x}{50}$ $x = 15$
3. When there's only one fraction, multiply by the denominator.	$\frac{2x}{5} = 10$ $2x = 50$ $x = 25$

If $\frac{4}{5}$ of the students in a certain class earned a passing grade,

$\frac{1}{6}$ failed, and the remaining 5 students had incompletes,

how many students were in the class?

(A) 20
(B) 30
(C) 50
(D) 100
(E) 150

SOLUTION NEXT PAGE

ARITHMETIC	QUIZ	FRACTIONS

NON-INTEGERS

If $\frac{4}{5}$ of the students in a certain class earned a passing grade,

$\frac{1}{6}$ failed, and the remaining 5 students had incompletes,

how many students were in the class?

SOLUTION

STEP 1 Separate the information from the question

$\frac{4}{5}$ = Passed $\frac{1}{6}$ = Failed 5 = Incomplete

STEP 2 Turn this into something you can use

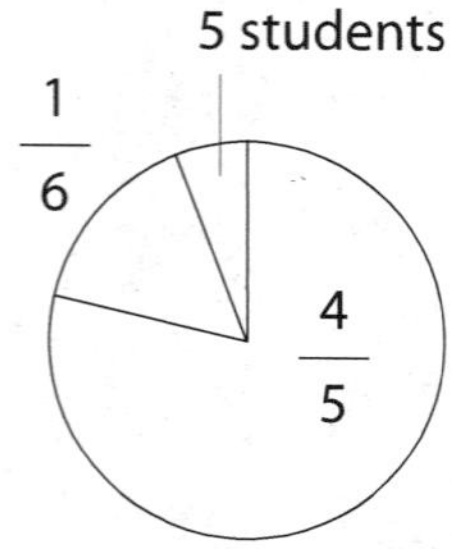

STEP 3 Use this to answer the question

$\frac{4}{5} + \frac{1}{6} = \frac{29}{30}$ of the students didn't get an Incomplete.

So $\frac{1}{30}$ of the students did get an Incomplete.

If $\frac{1}{30}$ of the class is 5 students, then set it up like this:

$$\frac{1}{30}\ class = 5$$

Multiply both sides by the denominator (30), and *class* = 150.

Answer is E

TOPIC	FRACTIONS — RECIPROCALS	ARITHMETIC
		NON-INTEGERS

Flip a fraction to find its reciprocal.

$$\frac{3}{4} \rightarrow \frac{4}{3} \text{ Reciprocal}$$

LIFELINE 25

When there's a 1 on top of a fraction, flip the fraction.

For example:

$$\frac{1}{\frac{7}{5}} \rightarrow \frac{5}{7}$$

QUIZ FRACTION — RECIPROCALS

$$\frac{1}{\frac{2}{5} + \frac{1}{1 + \frac{2}{3}}} =$$

(A) 1
(B) 2
(C) 3
(D) 4
(E) 5

SOLUTION NEXT PAGE

ARITHMETIC | QUIZ | FRACTIONS — RECIPROCALS

NON-INTEGERS

$$\frac{1}{\frac{2}{5}+\frac{1}{1+\frac{2}{3}}} =$$

SOLUTION

STEP 1 Separate the information from the question

Copy the fraction to your scratch paper

$$\frac{1}{\frac{2}{5}+\frac{1}{1+\frac{2}{3}}}$$

STEP 2 Turn this into something you can use

If the denominator is in pieces

$$\frac{1}{1+\frac{2}{3}}$$

Combine it before flipping

$$\frac{1}{\frac{5}{3}} \rightarrow \frac{3}{5}$$

Simplify the fraction, step by step, starting with the bottom

$$\frac{1}{1+\frac{2}{3}} = \frac{1}{\frac{5}{3}} = \frac{3}{5}$$

Add the two fractions

$$\frac{1}{\frac{2}{5}+\frac{3}{5}} = \frac{1}{\frac{1}{1}} = 1$$

STEP 3 Use this to answer the question

Answer is A

TOPIC	DECIMALS — ADDING, SUBTRACTING, MULTIPLYING	ARITHMETIC
		NON-INTEGERS

With these lifelines, you won't need a calculator.

LIFELINE 26

When adding or subtracting numbers with decimals, line the decimal points up.

Example:

$$\begin{array}{r} 6.00 \\ +\,.25 \\ \hline 6.25 \end{array}$$

LIFELINE 27

When multipling numbers with decimals, count the decimal spaces.

When both numbers have decimals — **.02 × .03**

1. Multiply the numbers without the decimals — 2 × 3 = 6
2. Count the decimal places of the original numbers — .02 × .03 = .0006 (← 2, ← 2, ← 4 spaces)
3. Move the decimal of your answer

When both numbers have zeros — **300 × 500**

1. Multiply the numbers without the zeros — 3 × 5 = 15
2. Count the zeros of the original numbers — 300 × 500 = 150,000 (→ 2, → 2, → 4 zeros)
3. Add the zeros to your answer

DRILL	DECIMALS — MULTIPLYING

Multiply, then count the decimals

1) .02 x .004 = .000008
2) 5 x .5 = 2.5
3) 70 x 40 = 2800
4) 800 x 5000 = 4000000
5) .03 x .05 = .0015

SOLUTION NEXT PAGE

ARITHMETIC

NON-INTEGERS

SOLUTION

DRILL **DECIMALS — MULTIPLYING**

1) .02 x .004 = .00008

2) 5 x .5 = 2.5

3) 70 x 40 = 2,800

4) 800 x 5000 = 4,000,000

5) .03 x .05 = .0015

LIFELINE 28

When multiplying one number with a decimal and one with zeros, move the decimals.

For example:	**7.5 × 20**
1. Move the decimal points of both numbers	7.5 × 20
a. the opposite direction	→ 1 ← 1
b. the same number of spaces	each one space
c. until both numbers are whole	75 × 2
2. Then multiply	75 × 2 = 150

DRILL **DECIMALS — MULTIPLYING**

Move the decimals, then multiply

1) 200 x .03 = ______

2) 80 x .4 = ______

3) 450 x .2 = ______

SOLUTION NEXT PAGE

DRILL — DECIMALS — MULTIPLYING

1) $200 \times .03 \rightarrow 2 \times 3 \rightarrow 6$
2) $80 \times .4 \rightarrow 8 \times 4 \rightarrow 32$
3) $450 \times .2 \rightarrow 45 \times 2 \rightarrow 90$

SOLUTION

TOPIC — DECIMALS — DIVIDING

Dividing decimals without a calculator is simple.

LIFELINE 29

When dividing numbers with decimals, move the decimals of both numbers to the right.

Example: When dividing numbers with decimals $\frac{.66}{.2}$

1. Move both decimal points to the right the same number of spaces until the one **divided by** is a whole number $\frac{.66}{.2} \rightarrow \frac{6.6}{2}$
2. Then divide the numbers $\frac{6.6}{2} = 3.3$

DRILL — DECIMALS — DIVIDING

1) $\frac{16}{.2}$

2) $\frac{.27}{.03}$

3) $\frac{1}{.004}$

SOLUTION NEXT PAGE

ARITHMETIC | **DRILL** | **DECIMALS — DIVIDING**

NON-INTEGERS

SOLUTION

1) $\frac{16}{.2} \rightarrow \frac{160}{2} \rightarrow 80$

2) $\frac{27}{.03} \rightarrow \frac{2700}{3} \rightarrow 900$

3) $\frac{1}{.004} \rightarrow \frac{1000}{4} \rightarrow 250$

QUIZ 1 | **DECIMALS — MULTIPLYING & DIVIDING** | **ARITHMETIC**

NON-INTEGERS

$$\frac{(.0014)(.32)(100)}{(.01)(.02)(.07)} =$$

(A) 3,200
(B) 320
(C) 32
(D) 3.2
(E) .32

SOLUTION NEXT PAGE

ARITHMETIC | QUIZ 1 | DECIMALS — MULTIPLYING AND DIVIDING

NON-INTEGERS

$$\frac{(.0014)(.32)(100)}{(.01)(.02)(.07)} =$$

(A) 3,200
(B) 320
(C) 32
(D) 3.2
(E) .32

SOLUTION

Remember
When multiplying fractions, cancel what you can.

STEP 1 Separate the information from the question

$$\frac{(.0014)(.32)(100)}{(.01)(.02)(.07)} \rightarrow \frac{.0014}{.01} \times \frac{.32}{.02} \times \frac{100}{.07}$$

STEP 2 Turn this into something you can use

Move the decimal points of the fractions divided

$$\frac{.14}{1} \times \frac{32}{2} \times \frac{10{,}000}{7}$$

Move the decimal points of the numbers multiplied (on the top: the .14 and the 10,000)

$$\frac{14.}{1} \times \frac{32}{2} \times \frac{100.}{7}$$

Cancel the 14 on top with the 2 and the 7 on the bottom.

$$\frac{\cancel{14}}{1} \times \frac{32}{\cancel{2}} \times \frac{100}{\cancel{7}}$$

STEP 3 Use this to answer the question

Multiply it out

$$\frac{1}{1} \times \frac{32}{1} \times \frac{100}{1} = 3{,}200$$

Answer is A

thinktankSCHOLASTIC

QUIZ 2 | **DECIMALS — MULTIPLYING & DIVIDING** | **ARITHMETIC**

NON-INTEGERS

$$\frac{(1.2)(.04)(.002)}{(.06)(.02)(.01)} =$$

(A) .008
(B) .08
(C) .8
(D) 8
(E) 80

C

SOLUTION NEXT PAGE

ARITHMETIC | QUIZ 2 | DECIMALS — DIVIDING

NON-INTEGERS

$$\frac{(1.2)(.04)(.002)}{(.06)(.02)(.01)} =$$

(A) .008
(B) .08
(C) .8
(D) 8
(E) 80

SOLUTION

STEP 1 Separate the information from the question

$$\frac{(1.2)(.04)(.002)}{(.06)(.02)(.01)} \rightarrow \frac{1.2}{.06} \times \frac{.04}{.02} \times \frac{.002}{.01}$$

STEP 2 Turn this into something you can use

The steps are easy. The hard part is staying organized.

Move the decimal points of the fractions divided

$$\frac{120}{6} \times \frac{4}{2} \times \frac{.2}{1}$$

Move the decimal points of the numbers multiplied (on top: the 120 and the .2)

$$\frac{\overleftarrow{12.}}{6} \times \frac{4}{2} \times \frac{\overrightarrow{2.}}{1}$$

Cancel the 12 and the 4 on top with the 6 and the 2 on the bottom.

$$\frac{\overset{2}{\cancel{12}}}{\cancel{6}} \times \frac{\overset{2}{\cancel{4}}}{\cancel{2}} \times \frac{2}{1}$$

STEP 3 Use this to answer the question

Multiply it all out

$$\frac{2}{1} \times \frac{2}{1} \times \frac{2}{1} = 8$$

Answer is D

TOPIC DECIMALS — SCIENTIFIC NOTATION | **ARITHMETIC**

NON-INTEGERS

Scientific notation refers to a number with a decimal times 10 to an exponent, such as 2.3×10^4

LIFELINE 30

With scientific notation, move the decimal the number of spaces shown by the exponent.

5.2×10^3 moves the decimal 3 spaces to the right = 5,200

6.3×10^{-3} moves the decimal 3 spaces to the left = .0063

4.5×10^0 moves the decimal 0 spaces = 4.5

Any number to the zero power equals 1.

$4.5 \times 10^0 = 4.5$

is the same as

$4.5 \times 1 = 4.5$

DRILL — DECIMALS — SCIENTIFIC NOTATION

Scientific Notation	Answer
1) $7.5 \times 10^2 =$	750
2) $10^2 =$	100
3) $3.4 \times 10^{-2} =$	.034
4) $6.1 \times 10^0 =$	6.1
5) $10^0 =$	1

QUIZ — DECIMALS — SCIENTIFIC NOTATION

If a manufacturer produced 3.6×10^8 units to fill 12,000 identical orders last year, how many units were in each order?

(A) 30
(B) 300
(C) 3,000
(D) 30,000
(E) 300,000

SOLUTION NEXT PAGE

ARITHMETIC | QUIZ | DECIMALS — SCIENTIFIC NOTATION

NON-INTEGERS

If a manufacturer produced 3.6×10^8 units to fill 12,000 identical orders last year, how many units were in each order?

(A) 30
(B) 300
(C) 3,000
(D) 30,000
(E) 300,000

SOLUTION

STEP 1 Separate the information from the question

3.6×10^8 units fill 12,000 identical orders

STEP 2 Turn this into something you can use

Turn these into a fraction $\frac{3.6 \times 10^8}{12{,}000}$

STEP 3 Use this to answer the question

Simplify the scientific notation $\frac{360{,}000{,}000}{12{,}000}$

Cancel the zeros $\frac{360{,}000}{12}$

Simplify the fraction by factoring

3
~~36~~ 10,000
\ /
~~360,000~~
~~12~~

Multiply the remaining factors $3 \times 10{,}000 = 30{,}000$

Answer is D

Remember
When factoring numbers with zeros, start with the zeros.

TOPIC | **PERCENTS**

ARITHMETIC

NON-INTEGERS

A percent is a fraction of 100. $20\% = \frac{20}{100}$

LIFELINE 31

When working with a percent, convert it to a decimal or fraction to make it easier.

Steps	**Example**
Turning a percent into a fraction	**40%**
1. Drop the % symbol	40
2. Place the number over 100	$\frac{40}{100}$
3. Reduce the fraction	$\frac{2}{5}$

Turning a percent into a decimal	**120%**	**13.5%**
1. Drop the % symbol	120	13.5
2. Move the decimal two spaces to the left.	1.20 ←·	.135 ←·
3. Drop any trailing zeros	1.2	.135

DRILL | **CONVERTING PERCENTS**

	As a fraction of 100	**As a decimal**	**As a reduced fraction**
75% equals	75/100	.75	3/4
150% equals	150/100	1.5	1 1/2
[0] **.5% equals**	5/100	.005	1/200

SOLUTION NEXT PAGE

ARITHMETIC | DRILL | CONVERTING PERCENTS

NON-INTEGERS

SOLUTION

	As a fraction of 100	As a decimal	As a reduced fraction
75% equals	$\frac{75}{100}$	.75	$\frac{3}{4}$
150% equals	$\frac{150}{100}$	1.5	$\frac{3}{2}$
.5% equals	$\frac{.5}{100}$	.005	$\frac{1}{200}$

Any time you get a percent, you can turn it into a fraction or decimal.

For example, to find 40% of 80, turn the 40% into either .4 or $\frac{2}{5}$ whichever is easier or works better in the question.

$.4 \times 80 = 4 \times 8 = 32$

I once had a student who turned every fraction and decimal into a percent. It worked for him, and he got all the answers right.

TOPIC FINDING THE PERCENT OF A PERCENT

ARITHMETIC

NON-INTEGERS

Finding a percent of a percent is like finding a fraction of a fraction: 50% of 50% = 25% as ½ of ½ = ¼

LIFELINE 32

To find the a percent of a percent, convert the percents to decimals or fractions, and multiply.

Example: If 20% of the employees get a promotion, and 30% of those promoted are transferred to a new office, what percent of the employees are transferred to a new office?

Steps	**Example**
To find the percent of a percent	**20% of 30%**
1. Convert to decimals	.2 of .3
2. Multiply	.2 × .3 = .06
3. Convert back to a percent	.06 = 6%

— Or —

1. Convert to fractions	$\frac{20}{100}$ of $\frac{30}{100}$
2. Reduce and multiply	$\frac{2}{10}$ of $\frac{3}{10}$
	$\frac{2}{10} \times \frac{3}{10} = \frac{6}{100}$
3. Convert back to a percent	$\frac{6}{100} = 6\%$

DRILL FINDING THE PERCENT OF A PERCENT

1) What is 25% of 60%

2) What is 70% of 40%

SOLUTION NEXT PAGE

ARITHMETIC | DRILL | FINDING THE PERCENT OF A PERCENT

NON-INTEGERS

SOLUTION

Steps | **Example**

1) What is 25% of 60%

1. Convert to fractions $\frac{25}{100}$ of $\frac{60}{100}$

2. Reduce and multiply $\frac{1}{4}$ of $\frac{3}{5}$

$$\frac{1}{4} \times \frac{3}{5} = \frac{3}{20}$$

3. Convert back to a percent $\frac{3}{20} = \frac{15}{100} = 15\%$

2) What is 70% of 40%

1. Convert to decimals .7 of .4

2. Multiply $.7 \times .4 = .28$

3. Convert back to a percent $.28 = 28\%$

QUIZ | **FINDING THE PERCENT OF A PERCENT** | **ARITHMETIC**

NON-INTEGERS

Of the 250 cars in a lot, 40% are convertible. If 50% of the convertibles are sold, what percent of the remaining cars are convertible?

(A) 20%
(B) 25%
(C) 30%
(D) 40%
(E) 50%

SOLUTION NEXT PAGE

ARITHMETIC | QUIZ | FINDING THE PERCENT OF A PERCENT

NON-INTEGERS

Of the 250 cars in a lot, 40% are convertible. If 50% of the convertibles are sold, what percent of the remaining cars are convertible?

(A) 20%
(B) 25%
(C) 30%
(D) 40%
(E) 50%

SOLUTION

STEP 1 Separate the information from the question

250 cars in a lot, 40% are convertible, and 50% of those are sold

STEP 2 Turn this into something you can use

Find the number of cars that are convertible	40% of 250
	$= .4 \times 250$
Move the decimals to multiply	$= 4 \times 25$
	$= 100$
Find the number of cars sold	50% of 100 = 50

Remember
When multiplying one number with a decimal and one with zeros, move the decimals.

STEP 3 Use this to answer the question

Of 100 convertibles, 50 are sold, so 50 remain.	50
Of 250 cars, 50 are sold, so 200 remain	200
Convert this to a percent	$\frac{50}{200} = \frac{25}{100} = 25\%$

Answer is B

TOPIC	FINDING THE PERCENT OF CHANGE	ARITHMETIC
		NON-INTEGERS

Finding the percent of change is common both on the GMAT and in business, especially finance.

LIFELINE 33

To find the percent of change, divide the amount of change by the original amount.

Example: If a price changes from $75 to $100.

Steps	**Example**
Find the percent of change	From 75 to 100
1. Find the amount of change	25
2. Divide by the **original** amount	$\frac{25}{75} = \frac{1}{3}$
3. Convert the denominator to 100 for %	$\frac{1}{3} = \frac{33}{100} = 33\%$

Find the percent of change	from 100 to 75
1. Find the amount of change	25
2. Divide by the **original** amount	$\frac{25}{100}$
3. Convert the denominator to 100 for %	$\frac{25}{100} = 25\%$

DRILL PERCENT OF CHANGE

Find the percents of change

1) from 50 to 40 $\frac{10}{50} = \frac{1}{5} = 20\%$

2) from 40 to 50 $\frac{10}{40} = .25 = \frac{1}{4} = 25\%$

SOLUTION NEXT PAGE

ARITHMETIC

NON-INTEGERS

SOLUTION

DRILL

PERCENT OF CHANGE

Steps	Example
Find the percent of change	from 50 to 40
1. Find the amount of change	10
2. Divide by the **original** amount	$\frac{10}{50} = \frac{1}{5}$
3. Convert the denominator to 100 for %	$\frac{1}{5} = \frac{20}{100} = 20\%$

Steps	Example
Find the percent of change	from 40 to 50
1. Find the amount of change	10
2. Divide by the **original** amount	$\frac{10}{40} = \frac{1}{4}$
3. Convert the denominator to 100 for %	$\frac{1}{4} = \frac{25}{100} = 25\%$

QUIZ | PERCENT OF CHANGE | ARITHMETIC

NON-INTEGERS

If on Monday the share price of a company stock opened at $100 and increased 20%, then on Tuesday decreased 20%, what was the closing price on Tuesday?

(A) $96
(B) $100
(C) $104
(D) $116
(E) $120

SOLUTION NEXT PAGE

ARITHMETIC | QUIZ | PERCENT OF CHANGE

NON-INTEGERS

If on Monday the share price of a company stock opened at $100 and increased 20%, then on Tuesday decreased 20%, what was the closing price on Tuesday?

(A) $96
(B) $100
(C) $104
(D) $116
(E) $120

SOLUTION

STEP 1 Separate the information from the question

The share price of a company stock opened at $100 and increased 20%, then decreased 20%

STEP 2 Turn this into something you can use

$100 → 20% increase → 20% decrease

STEP 3 Use this to answer the question

Starting price	$100
With a 20% increase	20% of $100 is $20
	$100 + $20 = $120
Then a 20% increase	20% of $120 is $24
	$120 – $24 = $96

Answer is A

TOPIC	PERCENT OF A GROUP	ARITHMETIC
		NON-INTEGERS

The percent of a group is intuitive. For example, 5 out of 10 is 50%. Use this intuitive grasp to work the trickier numbers.

LIFELINE 34

When the GMAT asks for the percent of a group, set it up as a fraction equal to $\frac{x}{100}$

Example: 4 of the 25 test takers earned scholarships. What percent of the test takers earned scholarships?

Steps	**Example**
What percent is 4 of 25?	
1. Place the number over the whole group	$\frac{4}{25}$
2. Set the fraction equal to	$\frac{4}{25} = \frac{x}{100}$
3. Solve for x	$\frac{4}{25} = \frac{16}{100}$ $x = 16$ 4 of 25 is 16%

DRILL — PERCENT OF A GROUP

1) What percent is 3 of 20?

2) What percent is 12 of 40?

SOLUTION NEXT PAGE

ARITHMETIC | DRILL | PERCENT OF A GROUP

NON-INTEGERS

SOLUTION

Steps | **Example**

1) What percent is 3 of 20?

1. Place the number over the whole group — $\frac{3}{20}$
2. Set the fraction equal to $\frac{x}{100}$ — $\frac{3}{20} = \frac{x}{100}$
3. Solve for x — $\frac{3}{20} = \frac{15}{100}$

$x = 15$

3 of 20 is 15%

2) What percent is 12 of 40?

1. Place the number over the whole group — $\frac{12}{40}$
2. Reduce the fraction — $\frac{12}{40} = \frac{3}{10}$
3. Set the fraction equal to $\frac{x}{100}$ — $\frac{3}{10} = \frac{x}{100}$
4. Solve for x — $x = 30$

12 of 40 is 30%

Remember
When one fraction has an x on top, give the fractions a common denominator.

Of the applicants for a certain job, 24 have graduate degrees. Of those who do not have graduate degrees, 16 have undergraduate degrees and the remaining 20 do not. What percent of the applicants have graduate degrees?

(A) 20
(B) 40
(C) 50
(D) 60
(E) 80

SOLUTION NEXT PAGE

ARITHMETIC | **QUIZ 1** | **PERCENT OF A GROUP**

NON-INTEGERS

Of the applicants for a certain job, 24 have graduate degrees. Of those who do not have graduate degrees, 16 have undergraduate degrees and the remaining 20 do not. What percent of the applicants have graduate degrees?

(A) 20
(B) 40
(C) 50
(D) 60
(E) 80

SOLUTION

STEP 1 Separate the information from the question

24 applicants have graduate degrees, 16 have undergraduate degrees and the remaining 20 have neither.

STEP 2 Turn this into something you can use

Of all the applicants:
- 24 have graduate degrees
- 16 have undergraduate degrees
- 20 have neither

For a total of 60 applicants

STEP 3 Use this to answer the question

1. Set the number with graduate degrees over the whole group $\frac{24}{60}$

2. Reduce the fraction $\frac{\not{24}}{\not{60}} = \frac{4}{10} = \frac{2}{5}$ (24 and 60 each divided by 6: 24 → 4, 60 → 10)

3. Set the fraction equal to $\frac{x}{100}$ $\qquad \frac{2}{5} = \frac{x}{100}$

4. Solve for x $\qquad \frac{(20)\,2}{(20)\,5} = \frac{40}{100}$

$$x = 40$$

Answer is B

thinktankSCHOLASTIC

QUIZ 2 PERCENT OF A GROUP

80% of the cars in the shop need oil changes. Among the cars needing oil changes, 33 also need radiator flushes and the remaining 15 do not. How many cars are in the shop?

(A) 60
(B) 80
(C) 100
(D) 120
(E) 140

SOLUTION NEXT PAGE

ARITHMETIC | **QUIZ 2** | **PERCENT OF A GROUP**

NON-INTEGERS

80% of the cars in the shop need oil changes. Among the cars needing oil changes, 33 also need radiator flushes and the remaining 15 do not. How many cars are in the shop?

(A) 60
(B) 80
(C) 100
(D) 120
(E) 140

SOLUTION

STEP 1 Separate the information from the question

80% of the cars in the shop need oil changes.
Of these, 33 need radiator flushes and 15 do not.

STEP 1 Separate the information from the question

80% = 33 + 15
80% = 48 cars

STEP 3 Use this to answer the question

1. Set the number needing oil changes over the whole group, which in this case is *x* $\frac{48}{x}$

2. Set the fraction equal to the percent over 100 $\frac{48}{x} = \frac{80}{100}$

3. Solve for *x*
   a. Reduce the fraction $\frac{48}{x} = \frac{8}{10}$
   b. Cross multiply $8x = 480$
   c. Divide both sides by 8 $x = 60$

Remember
When *x* is on the bottom, cross multiply.

Answer is A

TOPIC	COMPLETING THE RATIO	ARITHMETIC
		RATIOS

A ratio compares a group as a reduced fraction. For example, if there are 10 pens and 5 pencils, the fraction $\frac{10}{5}$ reduces to the ratio of 2 : 1.

LIFELINE 35

To complete a ratio, set it up as a fraction with *x* as the missing number.

$$\frac{\#\,dogs}{\#\,cats} = \frac{3}{2}$$

If the ratio of dogs to cats is 3 : 2, and there are 15 dogs, how many cats are there?

Steps	**Example**
1. Set up the equation	$\frac{\#\,dogs}{\#\,cats} = \frac{3}{2} = \frac{15}{x}$
2. Cross multiply	$3x = 30$
3. Solve for x	$x = 10$

Answer: With 15 dogs, there are 10 cats.

DRILL — COMPLETING THE RATIO

If the ratio of teachers to students is 2 : 25, and there are 150 students, how many teachers are there?

ARITHMETIC

RATIOS

SOLUTION

DRILL — COMPLETING THE RATIO

Steps	Example
1. Set up the equation	$\frac{\#\ teachers}{\#\ students} = \frac{2}{25} = \frac{x}{150}$
2. Cross multiply	$25x = 300$
3. Solve for x	$x = 12$

Answer: With 150 students, there are 12 teachers.

DETAIL — COMPLETING THE RATIO

Given a ratio, you can find a fraction of the group. For example, if the ratio of pens to pencils is 2 : 3,

then pens are $\frac{2}{5}$

of the total: $\frac{2}{2+3} = \frac{2}{5}$

TOPIC — NUMBER OF EACH ITEM

Given a ratio and a total, you can find the numbers of each item.

LIFELINE 36

When finding numbers in a ratio, set up the ratio as an equation.

If the ratio of cars to trucks is 2 : 5, and there are 35 vehicles total, how many cars and trucks are there?

Steps	Example
1. Start with the fraction	$\frac{\#\ cars}{\#\ trucks} = \frac{2x}{5x}$
2. The total is 35, so set up the equation to solve for x.	$2x + 5x = 35$ $7x = 35$ $x = 5$
3. Substitute the value of x in the original equation	$\frac{\#\ cars}{\#\ trucks} = \frac{2(5)}{5(5)} = \frac{10}{25}$

There are 10 cars and 25 trucks.

DRILL — NUMBER OF EACH ITEM

If the ratio of desks to chairs is 3 : 4, and there are 21 pieces of furniture total, how many desks and chairs are there?

QUIZ 1 — NUMBER OF EACH ITEM

In the annual budget for Cognos Corp., the dollar amounts allocated to utilities, mortgage, and staff are in the ratio of 4 : 5 : 6, respectively. If the total amount allocated to these categories is \$300,000, what is the amount allocated to utilities?

(A) \$ 20,000

(B) \$ 40,000

(C) \$ 60,000

(D) \$ 80,000

(E) \$100,000

SOLUTION NEXT PAGE

ARITHMETIC

RATIOS

DRILL — NUMBER OF EACH ITEM

If the ratio of desks to chairs is 3 : 4, and there are 21 pieces of furniture total, how many desks and chairs are there?

SOLUTION

1) Set up the equation: $3x + 4x = 21$

2) Solve for x: $7x = 21$

$x = 3$

3) Substitute the 3 for x in the equation: $3(3) + 4(3) = 21$

$9 + 12 = 21$

There are 9 desks and 12 chairs.

QUIZ 1 — NUMBER OF EACH ITEM

In the annual budget for Cognos Corp., the dollar amounts allocated to utilities, mortgage, and staff are in the ratio of 4 : 5 : 6, respectively. If the total amount allocated to these categories is \$300,000, what is the amount allocated to utilities?

(A) \$ 20,000

(B) \$ 40,000

(C) \$ 60,000

(D) \$ 80,000

(E) \$100,000

SOLUTION

STEP 1 Separate the information from the question

The dollar amounts are in the ratio of 4 : 5 : 6, for a total of 300,000

STEP 2 Turn this into something you can use

Steps	Example
1. Set up the equation	$4x + 5x + 6x = 300{,}000$
2. Solve for x:	$15x = 300{,}000$
	$x = 20{,}000$

STEP 3 Use this to answer the question

Utilities get $4x$, which is \$80,000

Answer is D

The ratio of the measures of the interior angles of a certain triangle is 2 : 3 : 4. What is the measure of the smallest angle?

(A) 20°

(B) 30°

(C) 40°

(D) 50°

(E) 60°

SOLUTION NEXT PAGE

ARITHMETIC | **QUIZ 2** | **NUMBER OF EACH ITEM**

RATIOS

The ratio of the measures of the interior angles of a certain triangle is 2 : 3 : 4. What is the measure of the smallest angle?

(A) 20°
(B) 30°
(C) 40°
(D) 50°
(E) 60°

SOLUTION

STEP 1 Separate the information from the question

The ratio of the angles is 2 : 3 : 4

STEP 2 Turn this into something you can use

The angles are in the ratio of 2 : 3 : 4, for a total of 180°

The angles of a triangle always total 180°. For more on this, see Geometry, p. 217

Steps	Example
1. Set up the equation	$2x + 3x + 4x = 180°$
2. Solve for x	$9x = 180°$
	$x = 20°$

STEP 3 Use this to answer the question

The smallest angle is $2x$, which is 40°.

Answer is C

DETAIL | **RATIOS / EXPRESSED AS A DECIMAL OR PERCENT**

A ratio can be expressed as a decimal.

A bike-to-skateboard ratio of 0.7 means 7 bikes per 10 skateboards,

because $0.7 = \frac{7}{10}$

A glove-to-helmet ratio of 2.5 means 5 gloves per 2 helmets,

because $2.5 = \frac{5}{2}$

A ratio can also be expressed as a percent:

A dog-to-pony ratio of 25% means 1 dog per 4 ponies,

because $25\% = \frac{1}{4}$

TOPIC	ADDING NUMBERS TO RATIOS	ARITHMETIC
		RATIOS

The GMAT introduces a ratio, changes the numbers, and asks for the new ratio. This is workable with the following step-by-step.

LIFELINE 37

When finding the new ratio, set up the old ratios as fractions equal to each other.

For example: The ratio of nickels to dimes is 5 : 4. After four of the dimes are changed into eight nickels, the new ratio is 7 : 3. How much money is there total?

Steps	**Example**
1. Set up the original equation with *x*'s, as a fraction	$\frac{\textit{Nickels}}{\textit{Dimes}} = \frac{5x}{4x}$
2. Adjust the fraction for the numbers that changed	$\frac{5x+8}{4x-4}$
3. Set this equal to the new ratio	$\frac{5x+8}{4x-4} = \frac{7}{3}$
4. Solve for *x*	
a. Cross multiply	$3(5x+8) = 7(4x-4)$
b. Distribute the 3 and the 7	$15x + 24 = 28x - 28$
c. Subtract 15*x* and add 28 to both sides	$52 = 13x$ $x = 4$
5. Substitute 4 for *x* in the original equation	$\frac{\textit{Nickels}}{\textit{Dimes}} = \frac{5x}{4x} = \frac{5(4)}{4(4)} = \frac{20}{16}$

20 nickels and 16 dimes is $2.60

DRILL	ADDING NUMBERS TO RATIOS

The numbers of juice boxes that Billy and Joey have are in the ratio of 5 : 3. After Billy gives Joey two of his boxes, the new ratio is 1 : 1. Before the gift, how many more juice boxes did Billy have?

SOLUTION NEXT PAGE

ARITHMETIC | DRILL | ADDING NUMBERS TO RATIOS

RATIOS

The numbers of juice boxes that Billy and Joey have are in the ratio of 5 : 3. After Billy gives Joey two of his boxes, the new ratio is 1 : 1. Before the gift, how many more juice boxes did Billy have?

SOLUTION

1. Set up the original equation with x's, as a fraction $\frac{Billy}{Joey} = \frac{5x}{3x}$
2. Adjust the fraction for the numbers that changed $\frac{5x-2}{3x+2}$
3. Set this equal to the new ratio $\frac{5x-2}{3x+2} = \frac{1}{1}$
4. Solve for x
   a. Cross multiply $5x - 2 = 3x + 2$
   b. Subtract $3x$ and add 2 to both sides $2x = 4$, $x = 2$
5. Substitute 4 for x in the original equation $\frac{Billy}{Joey} = \frac{5x}{3x} = \frac{5(2)}{3(2)} = \frac{10}{6}$

Billy had 4 more than Joey.

In a flag-football game, the numbers of red team players to blue team players is in the ratio of 3 : 2. However, the blue team is still winning, so 4 blue-team members switched to the red team, for a new red team : blue team ratio of 4 : 1. How many players were on the red team after the switch?

(A) 8
(B) 12
(C) 16
(D) 20
(E) 24

SOLUTION NEXT PAGE

ARITHMETIC | QUIZ | RATIOS / ADDING NUMBERS TO RATIOS

RATIOS

In a flag-football game, the numbers of red team players to blue team players is in the ratio of 3 : 2. However, the blue team is still winning, so 4 blue-team members switched to the red team, for a new red team : blue team ratio of 4 : 1. How many players were on the red team after the switch?

SOLUTION

STEP 1 Separate the information from the question

The original red : blue ratio is 3:2.
4 members switched from blue to red.
The new red : blue ratio is 4:1.

STEP 2 Turn this into something you can use

1. Set up the original equation with x's, as a fraction $\frac{Red}{Blue} = \frac{3x}{2x}$
2. Adjust the fraction for the numbers that changed $\frac{3x+4}{2x-4}$
3. Set this equal to the new ratio $\frac{3x+4}{2x-4} = \frac{4}{1}$
4. Solve for x
   a. Cross multiply $3x + 4 = 4(2x - 4)$
   b. Distribute the 4 $3x + 4 = 8x - 16$
   c. Subtract 3x and add 16 to both sides $20 = 5x$, $x = 4$
5. Substitute 4 for x in the original equation $\frac{Red}{Blue} = \frac{3x}{2x} = \frac{3(4)}{2(4)} = \frac{12}{8}$

STEP 3 Use this to answer the question

The red team started with 12 players and gained 4 during the switch, for a total of 16 players after the switch.

Answer is C

TOPIC **COMBINED RATIOS** — ARITHMETIC

RATIOS

Ratios can be combined when they have something in common.

LIFELINE 38

When combining ratios, stack the part that they have in common.

Example: If the ratio of kids to dogs is 5 : 2, and the ratio of dogs to Frisbees is 3 : 1, what is the ratio of kids to Frisbees?

To combine the ratios — kids to dogs is 5 : 2. dogs to Frisbees is 3 : 1

Steps	**Example**
1. Write out the first ratio	k : d 5 : 2
2. Add the second ratio, stacking the dogs up in a column	k : d : f 5 : 2 : 3 : 1
3. Underneath the dog column, write the lowest number of dogs that could be in both ratios	k : d : f 5 : 2 3 : 1 6
4. Cover up the first ratio. If the 3 becomes 6, the 1 becomes 2	k : d : f [ ] 3 : 1 6 : 2
5. Cover up the second ratio. If the 2 becomes a 6, the 5 becomes 15	k : d : f 5 : 2 [ ] 15 : 6 : 2
The combined ratio is 15 : 6 : 2, and the ratio of kids to Frisbees is 15 : 2	k : d : f 5 : 2 3 : 1 15 : 6 : 2

Finding the common multiple in ratios is like finding the common denominator of added fractions:

$$\frac{1}{3} + \frac{1}{2} = \frac{2}{6} + \frac{3}{6}$$

To find the missing number, treat the ratio like a fraction:

$\frac{3}{1} = \frac{6}{x}$, making $x = 2$

ARITHMETIC | **DRILL** | **COMBINED RATIOS**

RATIOS

The ratio of boats to scuba divers is 2 : 15
The ratio of scuba divers to tanks is 2 : 3
What's the ratio of tanks to boats?

SOLUTION

To find the ratio of tanks to boats

Steps	Example
1. Write out the first ratio	b : s 2 :15
2. Add the second ratio, stacking the scuba divers in a column	b : s : t 2 : 15 2 : 3
3. Underneath the scuba diver column, write the lowest number of divers that could be in both columns	b : s : t 2 : 15 2 : 3 30
4. Cover up the first ratio. If the 2 becomes 30, the 3 becomes 45	b : s : t 2 : 3 30 : 45
4. Cover up the second ratio. If the 15 becomes 30, the 2 becomes 4	b : s : t 2 : 15 4 : 30 : 45
The combined ratio is 4 : 30 : 45, and the ratio of tanks to boats is 45 : 4	b : s : t 2 : 15 2 : 3 4 : 30 : 45

To make a pizza, a certain pizzeria uses a ratio of sauce to water of 3 cups to 4 cups, respectively, and a ratio of water to flour of 6 cups to 11 cups, respectively. If the pizzeria used 27 cups of sauce on a certain day, how many cups of flour did it use?

(A) 22

(B) 33

(C) 44

(D) 55

(E) 66

SOLUTION NEXT PAGE

ARITHMETIC	QUIZ	COMBINED RATIOS

RATIOS

To make a pizza, a certain pizzeria uses a ratio of sauce to water of 3 cups to 4 cups, respectively, and a ratio of water to flour of 6 cups to 11 cups, respectively. If the pizzeria used 27 cups of sauce on a certain day, how many cups of flour did it use?

(A) 22
(B) 33
(C) 44
(D) 55
(E) 66

SOLUTION

STEP 1 Separate the information from the question

The ratio of sauce to water is 3 : 4, the ratio of water to flour is 6 : 11, and they used 27 cups of sauce.

STEP 2 Turn this into something you can use

> A common mistake is using 24 instead of 12 for the water ratio.
>
> Though 24 works, 12 is the lowest common multiple of 4 and 6.

Combine the ratios

s	w	f
3	4	
	6	11
9	12	22

STEP 3 Use this to answer the question

The restaurant used 27 cups of sauce, so multiply the combined ratio by 3 to get a sauce number of 27.

s : w : f
3(9 : 12 : 22)

The pizzeria used 66 cups of flour.

s : w : f
27 : 36 : 66

Answer is E

TOPIC	NEGATIVE EXPONENTS	ARITHMETIC
		BASIC EXPONENTS

Every number — 5

besides having an invisible 1 underneath — $\frac{5}{1}$

also has an invisible 1 as an exponent — $5^1 = \frac{5^1}{1} = 5$

LIFELINE 39

When a number has a negative exponent, place a 1 on top.

Steps	**Example**
A number with a **positive** exponent	5^2
has a 1 underneath it	$\frac{5^2}{1} = \frac{25}{1} = 25$
A number with a **negative** exponent	5^{-2}
has a 1 on top	$5^{-2} = \frac{1}{5^2} = \frac{1}{25}$

DRILL — NEGATIVE EXPONENTS

1) $12^1 =$

2) $15^{-1} =$

3) $5^3 =$

4) $2^{-3} =$

5) $5^0 =$

SOLUTION NEXT PAGE

ARITHMETIC

DRILL — **NEGATIVE EXPONENTS**

BASIC EXPONENTS

1) $12^1 = 12$

2) $15^{-1} = \dfrac{1}{15}$

SOLUTION

3) $5^3 = 5 \times 5 \times 5 = 125$

4) $2^{-3} = \dfrac{1}{2^3} = \dfrac{1}{2 \times 2 \times 2} = \dfrac{1}{8}$

5) $5^0 = 1$

> Any number to the 0 power equals 1

LIFELINE 40

When multiplying numbers with exponents, add the exponents.

Example:	$(x^3)(x^2)$
x^3 is x times itself 3 times	$x^3 = (x)(x)(x)$
x^2 is x times itself twice	$x^2 = (x)(x)$
Multiply them together by adding the exponents	$(x)(x)(x)*(x)(x) = x^{3+2} = x^5$

LIFELINE 41

When dividing numbers with exponents, subtract the exponents.

Example:	$\dfrac{x^3}{x^2}$
Expand the x^3 and the x^2	$\dfrac{(x)(x)(x)}{(x)(x)}$
The x's cancel	$\dfrac{(x)\cancel{(x)}\cancel{(x)}}{\cancel{(x)}\cancel{(x)}}$

Divide them by subtracting the exponents $x^{3-2} = x^1$ or x

This is why any number to the 0 power equals 1. $\dfrac{x^4}{x^4} = x^{4-4} = x^0 = 1$

> Even $0^0 = 1$

DRILL	MULTIPLYING & DIVIDING	ARITHMETIC

BASIC EXPONENTS

1) $(x^4)(x^2)$

2) $\dfrac{x^5}{x^3}$

3) $\dfrac{x^3}{x^5}$

4) $\dfrac{x^6}{x^6}$

5) If $y = 2$, then $\dfrac{2^{5y}}{2^{4y}} =$

SOLUTION

Drill Answer

1) x^6

2) x^2

3) $\dfrac{1}{x^2}$ or x^{-2}

4) $x^{6-6} = x^0 = 1$

5) $2^{5y-4y} = 2^y = 2^2 = 4$

QUIZ	MULTIPLYING & DIVIDING

If $y = \dfrac{1}{2}$, what is the value of $\dfrac{3^{2x+2y}}{3^{2x-2y}}$?

(A) 0
(B) 1
(C) 2
(D) 3
(E) 9

SOLUTION NEXT PAGE

ARITHMETIC

QUIZ MULTIPLYING & DIVIDING

BASIC EXPONENTS

If $y = \frac{1}{2}$, what is the value of $\frac{3^{2x+2y}}{3^{2x-2y}}$?

SOLUTION

STEP 1 Separate the information from the question

$y = \frac{1}{2}$ and $\frac{3^{2x+2y}}{3^{2x-2y}}$

STEP 2 Turn this into something you can use

Steps	Example
Subtract the exponents to divide the numbers	$\frac{3^{2x+2y}}{3^{2x-2y}} \rightarrow 3^{2x+2y-(2x-2y)}$
Distribute the negative among the terms in the brackets	$3^{2x+2y-2x+2y}$
Simplify the exponent: the x's cancel and the y's become $4y$	3^{4y}

This works exactly the same way as

$\frac{x^5}{x^3} \rightarrow x^{5-3} \rightarrow x^2$

STEP 3 Use this to answer the question

Because $y = \frac{1}{2}$, $4y = 2$ $\quad 3^2 = 9$

Answer is E

DETAIL EXPONENTS

The exponent only multiplies the term that it's attached to.

With -5^2, the exponent multiplies the 5, but not the negative sign:

-5^2

$-(5 \times 5)$

-25

With $(-5)^2$, the exponent multiplies the entire bracket, including the negative sign:

$(-5)^2$

$(-5) \times (-5)$

25

Think of the exponent as stuck to the term with chewing gum.

TOPIC | **EXPONENTS OF EXPONENTS** | **ARITHMETIC**

BASIC EXPONENTS

An **exponent of an exponent** means that the entire expression is multiplied by itself that number of times.

LIFELINE 42

When taking the exponent of an exponent, multiply the exponents.

Example: $(x^2)^4$

$(x^2)^4$ means x^2 is multiplied by itself 4 times: $(x^2)(x^2)(x^2)(x^2)$

$x^{2+2+2+2} = x^{4\times 2} = x^8$

DRILL | **EXPONENTS OF EXPONENTS**

Question:	**Answer:**
1) $(x^3)^5$	$x^{3\times 5} = x^{15}$
2) $(x^2)^{-4}$	$x^{(2)(-4)} = x^{-8}$
3) $(x^{-3})^{-5}$	$x^{(-3)(-5)} = x^{15}$

A negative times a negative is a positive, even in this case.

QUIZ | **EXPONENTS OF EXPONENTS**

If $x \neq 0$, then $\dfrac{x^2(x^3)^4}{x^7} =$

(A) x^{-1}
(B) x^2
(C) x^3
(D) x^5
(E) x^7

SOLUTION NEXT PAGE

ARITHMETIC	QUIZ	EXPONENTS OF EXPONENTS

BASIC EXPONENTS

If $x \neq 0$, then $\dfrac{x^2(x^3)^4}{x^7} =$

(A) x^{-1}
(B) x^2
(C) x^3
(D) x^5
(E) x^7

SOLUTION

STEP 1 Separate the information from the question

$$\frac{x^2(x^3)^4}{x^7}$$

STEP 2 Turn this into something you can use

1. Work the exponent of the exponent first $\dfrac{x^2x^{12}}{x^7}$

2. Multiply the exponents by adding them $\dfrac{x^{14}}{x^7}$

STEP 3 Use this to answer the question

3. Divide the exponents by subtracting them $x^{14-7} = x^7$

Answer is E

TOPIC FACTORING & CONVERTING

ARITHMETIC

ADVANCED EXPONENTS

Exponents with numbers are similar to exponents with *x*'s, only with a few key differences.

LIFELINE 43

When multiplying or dividing numbers with different exponents, add the exponents.

Steps	Example
1. When the **terms** are the same	$(2^3)(2^{10})$
2. Add the exponents	2^{3+10}
3. Keep the term the same	2^{13}

This works exactly the same as

$(x^2)(x^3) \rightarrow x^{2+3} \rightarrow x^5$

LIFELINE 44

When multiplying or dividing numbers with the same exponent, keep the exponent.

1. When the **exponents** are the same	$(9^3)(5^3)$
2. Multiply the terms	$(9 \times 5)^3$
3. Keep the exponent the same	45^3

The advanced exponent is is in my list of Top 5 GMAT question types.

DETAIL FACTORING & CONVERTING

When the **terms** and the **exponents** are the same	$(4^5)(4^5)$
Either multiply the terms, keeping the exponent the same	$(4 \times 4)^5 = 16^5$
Or add the exponents, keeping the term the same	$4^{5+5} = 4^{10}$
But **not both.**	16^{10} is wrong.

ARITHMETIC | **DETAIL** | **FACTORING & CONVERTING**

ADVANCED EXPONENTS

A number with an exponent can be factored in two ways.

Steps	Example
1. To factor the exponent	12^5
2. Separate the exponent	12^{2+3}
3. Place the separated exponents on each term, keeping the term the same	$12^2 \times 12^3$

This works exactly the same way as

$x^5 \rightarrow x^{2+3} \rightarrow (x^2)(x^3)$

Steps	Example
1. To factor the term	12^5
2. Factor the term separately	$(4 \times 3)^5$
3. Place the exponent on each factor, keeping the exponent the same	$4^5 \times 3^5$

Factoring both the term and the exponent is a common mistake.

DETAIL | **FACTORING EXPONENTS**

8^7 can be factored in two ways	8^7
Either separate the exponent, keeping the term the same	$8^4 \times 8^3$
Or factor the term, keeping the exponent the same	$4^7 \times 2^7$
But **not both**	$4^4 \times 2^3$ is wrong

Fill in the *x*:

1) $(2^7)(2^3) = 2^x$

2) $(5^3)(4^3) = x^3$

3) $16^5 = 2^x$

4) $(6^6)(2^3) = x^3$

5) $(16^6)(4^4) = 4^x$

SOLUTION NEXT PAGE

ARITHMETIC | **DRILL** | **FACTORING & CONVERTING**

ADVANCED EXPONENTS

Fill in the x:

> This works the same way as:
>
> $(x^2)(x^3) \rightarrow x^{2+3} \rightarrow x^5$

1) $(2^7)(2^3) = 2^{7+3} = 2^{10}$

2) $(5^3)(4^3) = (5 \times 4)^3 = 20^3$

3) $16^5 = (2^4)^5 = 2^{20}$

SOLUTION

3) This one can also be solved by factoring and multiplying:

16^5
/ \
4^5 4^5
/ \ / \
2^5 2^5 2^5 2^5
\ / \ /
2^{10} 2^{10}
\ /
2^{20}

> To get an answer with an exponent of 3, convert the 6^6 to 36^3

4) $(6^6)(2^3) =$

$6^6 \times 2^3$
/ \
6^3 $6^3 \times 2^3$
\ /
$36^3 \times 2^3$
\ /
72^3

> To get an answer with an term of 4, convert the 16^6 to 4^{12}

5) $(16^6)(4^4) =$

$16^6 \times 4^4$
/ \
4^6 $4^6 \times 4^4$
\ /
$4^{12} \times 4^4$
\ /
4^{16}

QUIZ 1 | FACTORING & CONVERTING

ARITHMETIC

ADVANCED EXPONENTS

If $5^{11} \times 4^{5} = 5 \times 10^{k}$, what is the value of k?

(A) 10

(B) 12

(C) 16

(D) 20

(E) 50

SOLUTION NEXT PAGE

ARITHMETIC | QUIZ 1 | FACTORING & CONVERTING

ADVANCED EXPONENTS

If $5^{11} \times 4^5 = 5 \times 10^k$, what is the value of k ?

(A) 10
(B) 12
(C) 16
(D) 20
(E) 50

SOLUTION

STEP 1 Separate the information from the question

$5^{11} \times 4^5 = 5 \times 10^k$

STEP 2 Turn this into something you can use

To isolate the 10^k $\quad 5^{11} \times 4^5 = 5 \times 10^k$

Divide both sides by 5: $\quad \frac{5^{11} \times 4^5}{5} = \frac{5 \times 10^k}{5}$

5's on both sides cancel $\quad 5^{10} \times 4^5 = 10^k$

To get a term of 10 on the left, convert the 4^5 into 2^{10} $\quad 5^{10} \times 4^5 \rightarrow 5^{10} \times 2^5\ 2^5$

And combine the 5 and the 2 $\quad 5^{10} \times 2^{10}$

$10^{10} = 10^k$

$k = 10$

STEP 3 Use this to answer the question

For $10^{10} = 10^k$, the exponents have to be the same, meaning $10 = k$

Answer is A

The question comes up: in this fraction ... $\dfrac{5^{11} \times 4^5}{5}$

Why wouldn't the 5 cancel from both the 5^{11} and the 4^5 ?

Because the terms are **multiplied**, $\dfrac{5^{11} \times 4^5}{5}$

the 5 cancels with only one term on the top. $\dfrac{\cancel{5^{11}} \times 4^5}{\cancel{5}} = 5^{10} \times 4^5$

Here's the difference: $\dfrac{2x+4}{2}$

Because the terms are **added** $\dfrac{2(x+2)}{2}$

Factor out the 2 and cancel it: $\dfrac{\cancel{2}(x+2)}{\cancel{2}} = x+2$

And if the terms were multiplied: $\dfrac{(2x)(4)}{2}$

The fraction could be written like this $\dfrac{(2)(x)(4)}{2}$

And the 2 on the bottom cancels with only one term on the top. $\dfrac{\cancel{(2)}(x)(4)}{\cancel{2}}$

QUIZ 2 FACTORING & CONVERTING

Which of the following is equal to $3^3 + 3^3 + 3^3 + 3^3 + 3^3 + 3^3 + 3^3 + 3^3 + 3^3$?

(A) 3^4
(B) 3^5
(C) 3^6
(D) 3^9
(E) 3^{27}

SOLUTION NEXT PAGE

ARITHMETIC | QUIZ 2 | FACTORING & CONVERTING

ADVANCED EXPONENTS

Which of the following is equal to $3^3 + 3^3 + 3^3 + 3^3 + 3^3 + 3^3 + 3^3 + 3^3 + 3^3$?

(A) 3^4
(B) 3^5
(C) 3^6
(D) 3^9
(E) 3^{27}

SOLUTION

STEP 1 Separate the information from the question

$$3^3 + 3^3 + 3^3 + 3^3 + 3^3 + 3^3 + 3^3 + 3^3 + 3^3$$

STEP 2 Turn this into something you can use

You can multiply and divide exponents,
but you can't add and subtract them.

However, you can count them.

$$3^3 + 3^3 + 3^3 + 3^3 + 3^3 + 3^3 + 3^3 + 3^3 + 3^3 = 9(3^3)$$

STEP 3 Use this to answer the question

$9(3^3)$ doesn't match any of the answer choices,
but you can convert it.

$$9(3^3) = 3^2(3^3) = 3^5$$

Exponents and algebra go hand-in-hand.
We revisit exponents in Algebra, p.169

Answer is B

TOPIC SIMPLIFYING SQUARE ROOTS — **ARITHMETIC**

RADICALS

1. **A square root has two answers:** $x^2 = 16$
   One positive and one negative. $x = 4$ or -4
   x has one answer, and you don't know which it is.

2. **If the radical is already in place,** $\sqrt{25} = 5$
   just use the positive answer.

3. **The square root can only be** $\sqrt{-1}$ is not a real number
   on a positive number.
   This is because no number can be
   multiplied by itself and stay negative.

LIFELINE 45

When simplifying a square root, factor it out.

Example: Factoring a square root

$\sqrt{48}$
/ \
$\sqrt{16}$ $\sqrt{3}$
/ \
$\sqrt{4}$ $\sqrt{4}$

The two $\sqrt{4}$'s become a single 4. However, the $\sqrt{3}$ doesn't have a match, so it stays under the radical.

The numbers are all multiplied, and the answer is $4\sqrt{3}$

DRILL SIMPLIFYING SQUARE ROOTS

1. $\sqrt{225}$ 2. $\sqrt{125}$ 3. $\sqrt{128}$

SOLUTION NEXT PAGE

ARITHMETIC

RADICALS

SOLUTION

DRILL — SIMPLIFYING SQUARE ROOTS

1. $\sqrt{225}$

   / \

   $\sqrt{25}$ $\sqrt{9}$

   | |

   5 3

   \ /

   15

2. $\sqrt{125}$

   / \

   $\sqrt{25}$ $\sqrt{5}$

   | |

   5 |

   \ |

   $5\sqrt{5}$

3. $\sqrt{128}$

   / \

   $\sqrt{64}$ $\sqrt{2}$

   | |

   8 |

   \ |

   $8\sqrt{2}$

DETAIL — SIMPLIFYING SQUARE ROOTS

Any number to the ½ power is the square root of that number.

$16^{1/2} = \sqrt{16} = 4$

For example, $25^{.5} = \sqrt{25} = 5$

TOPIC | SIMPLIFYING CUBE ROOTS

ARITHMETIC

RADICALS

1. **The cube root has one answer** $\sqrt[3]{8} = 2$

2. **The cube root can be on a positive or a negative number.** A negative number multiplied by itself three times remains negative. $\sqrt[3]{-8} = -2$

LIFELINE 46

When simplifying a cube root, look for triple factors.

Example:

$$\sqrt[3]{40}$$

$$\sqrt[3]{4} \qquad \sqrt[3]{10}$$

$$\boxed{\sqrt[3]{2}\ \sqrt[3]{2}\ \sqrt[3]{2}}\ \sqrt[3]{5}$$

$$2$$

The three $\sqrt[3]{2}$'s become 2, and the $\sqrt[3]{5}$ stays.

$$2\sqrt[3]{5}$$

Like the square root, these are all multiplied.

DRILL | SIMPLIFYING CUBE ROOTS

1) $\sqrt[3]{27}$

2) $\sqrt[3]{-64}$

3) $\sqrt[3]{128}$

SOLUTION NEXT PAGE

ARITHMETIC | **DRILL** | **SIMPLIFYING ROOTS / CUBE ROOTS**

RADICALS

SOLUTION

Drill Answer

1)

$\sqrt[3]{27}$

$\sqrt[3]{9}$

$\sqrt[3]{3}\ \sqrt[3]{3}\ \sqrt[3]{3}$

3

2)

$\sqrt[3]{-64}$

$\sqrt[3]{16}$

$-\ \sqrt[3]{4}\ \sqrt[3]{4}\ \sqrt[3]{4}$

-4

3)

$\sqrt[3]{128}$

$\sqrt[3]{64}\ \ \sqrt[3]{2}$

$4\sqrt[3]{2}$

TOPIC MULTIPLYING & DIVIDING — **ARITHMETIC**

RADICALS

Radicals can be multiplied and divided $\sqrt{2} \times \sqrt{3} = \sqrt{6}$

But not added or subtracted $\sqrt{2} + \sqrt{3}$ stays the same

LIFELINE 47

When multiplying or dividing numbers with radicals, ignore the radicals.

Steps	**Example**
Multipling radicals	$\sqrt{5} \times \sqrt{3}$
1. Ignore the radicals	5×3
2. Multiply the numbers	$5 \times 3 = 15$
3. Replace the radical	$15 \rightarrow \sqrt{15}$

Dividing Radicals	$\frac{\sqrt{10}}{\sqrt{2}}$
1. Ignore the radicals	$\frac{10}{2}$
2. Divide the numbers	$\frac{10}{2} = 5$
3. Replace the radical	$5 \rightarrow \sqrt{5}$

Multiplying Numbers with Radicals	$2\sqrt{3} \times 4\sqrt{5} =$
1. Multiply the numbers first	$2 \times 4 = 8$
2. Multiply the radicals second	$\sqrt{3} \times \sqrt{5} = \sqrt{15}$
3. Multiply the two results	$8 \times \sqrt{15} \rightarrow 8\sqrt{15}$

DRILL MULTIPLYING & DIVIDING

1) $3\sqrt{5} \times 7\sqrt{2}$ 2) $\frac{\sqrt{40}}{\sqrt{10}}$ 3) $3 \times 4\sqrt{5}$ 4) $\frac{8\sqrt{15}}{2\sqrt{5}}$

SOLUTION NEXT PAGE

ARITHMETIC	DRILL	MULTIPLYING & DIVIDING

RADICALS

SOLUTION

1)	$3\sqrt{5} \times 7\sqrt{2}$
Multiply the radicals first and the numbers second	$3 \times 7 = 21$ and $\sqrt{5} \times \sqrt{2} = \sqrt{10}$
Combine the results	$21 \times \sqrt{10} \rightarrow 21\sqrt{10}$
2)	$\dfrac{\sqrt{40}}{\sqrt{10}}$
Divide the numbers without the radicals	$\dfrac{40}{10} = 4$
Replace the radical	$4 \rightarrow \sqrt{4}$
Simplify the radical	$\sqrt{4} = 2$
3)	$3 \times 4\sqrt{5}$
Multiply the numbers without the radical and combine the results	$3 \times 4 \times \sqrt{5}$ $12 \times \sqrt{5} \rightarrow 12\sqrt{5}$
4)	$\dfrac{8\sqrt{15}}{2\sqrt{5}}$
Divide the numbers first and the radicals second	$\dfrac{8}{2} \times \dfrac{\sqrt{15}}{\sqrt{5}}$
Combine the results	$4 \times \sqrt{3} \rightarrow 4\sqrt{3}$

TOPIC | **ESTIMATING ROOTS** | **ARITHMETIC** | **RADICALS**

On the GMAT, a ballpark estimate is always good enough.

LIFELINE 48

When estimating a root, find the whole numbers just above and below the root.

Steps	**Example**
To estimate the root	$\sqrt{14}$
1. Find the whole number just above the root	$\sqrt{16} = 4$
2. Find the whole number just below the root	$\sqrt{9} = 3$
3. The root in question is between the two	$\sqrt{14} \approx 3.5$

DRILL | **ESTIMATING ROOTS**

$\sqrt{10}$ is closest to which of the following?

(A) 2.9
(B) 3.0
(C) 3.1
(D) 3.9
(E) 4.0

SOLUTION NEXT PAGE

ARITHMETIC

RADICALS

DRILL ESTIMATING ROOTS

$\sqrt{10}$ is closest to which of the following?

(A) 2.9
(B) 3.0
(C) 3.1
(D) 3.9
(E) 4.0

SOLUTION

Answer: $\sqrt{10}$ is in between $\sqrt{9}$ and $\sqrt{16}$, which equal 3 and 4, respectively. However, it is closer to $\sqrt{9}$ than to $\sqrt{16}$, making the correct answer (C), 3.1.

DETAIL ESTIMATING ROOTS

Always multiply first and estimate last.

Multiplying then estimating for a correct answer: $\sqrt{3} \times \sqrt{7} = \sqrt{21} \approx 4.5$

is both easier and more accurate than estimating then multiplying

for a wrong answer: $\sqrt{3} \times \sqrt{7} \approx 1.5 \times 2.5 \approx 3.75$

QUIZ | ESTIMATING ROOTS | ARITHMETIC

RADICALS

Of the following, which is closest to $(\sqrt{3}+\sqrt{7})^2$?

(A) 10

(B) 17

(C) 19

(D) 58

(E) 100

SOLUTION NEXT PAGE

ARITHMETIC

QUIZ — ESTIMATING ROOTS

RADICALS

Of the following, which is closest to $(\sqrt{3}+\sqrt{7})^2$?

(A) 10
(B) 17
(C) 19
(D) 58
(E) 100

SOLUTION

STEP 1 Separate the information from the question

$$(\sqrt{3}+\sqrt{7})^2$$

STEP 2 Turn this into something you can use

Foil the expressions $(\sqrt{3}+\sqrt{7})(\sqrt{3}+\sqrt{7})$

$3+2\sqrt{21}+7$

$=10+2\sqrt{21}$

> Did you distribute the exponent –
>
> $(\sqrt{3}^2+\sqrt{7}^2)$
>
> instead of foiling?
>
> $(\sqrt{3}+\sqrt{7})(\sqrt{3}+\sqrt{7})$
>
> This is a common mistake. The exponent squares the whole bracket.
>
> More on foiling in Algebra, p. 159

STEP 3 Use this to answer the question

1. Estimate $\sqrt{21}$
   as between $\sqrt{16}=4$ and $\sqrt{25}=5$ ≈ 4.5
2. Replace $\sqrt{21}$ with 4.5 in the expression $=10+2(4.5)$
   $=19$

Answer is C

TOPIC | **COMPLETING ROOTS** | **ARITHMETIC**

RADICALS

A square or cube root is **incomplete** if it's missing a factor.
Factor it out to find what's missing.

LIFELINE 49

When completing a radical, factor it out.

Example: If $\sqrt{2x}$ is an integer ...

x has to provide a match for the $\sqrt{2}$

```
   √2x
   /  \
 √2    √x
   \  /
    2
```

$$\sqrt{x} = \sqrt{2}$$
$$x = 2$$

$\sqrt{x}$ could also equal $\sqrt{8}$, making $\sqrt{2x} = \sqrt{16}$, which is an integer.

However, it could not equal $\sqrt{6}$, because $\sqrt{2x} = \sqrt{12}$ is not an integer.

Steps	Example
To complete the root	$\sqrt{150x}$ = an integer
1. Without x, factor the number to its primes	$\sqrt{150x}$ → $\sqrt{15}$, $\sqrt{10}$ → $\sqrt{3}$, [$\sqrt{5}$ $\sqrt{5}$], $\sqrt{2}$ → *, 5, *
2. Figure out what is missing to complete the root	$\sqrt{3}$ and $\sqrt{2}$
3. The x provides the missing factors	$\sqrt{x} = \sqrt{3} \times \sqrt{2}$ $\sqrt{x} = \sqrt{6}$ $x = 6$

DRILL | **COMPLETING ROOTS**

Find the lowest possible values of x:

1) $\sqrt{45x}$ = an integer

2) $\sqrt{60x}$ = an integer

3) $\sqrt[3]{72x}$ = an integer

SOLUTION NEXT PAGE

ARITHMETIC | DRILL | COMPLETING ROOTS

RADICALS

SOLUTION

1) To complete the square root $\sqrt{45x}$ = an integer

1. Without the x, factor the number to its primes

$\sqrt{9}\ \sqrt{5}$

$\boxed{\sqrt{3}\ \sqrt{3}}$

2. Figure out what is missing to complete the root

3 *

3. x provides the missing factors

$\sqrt{x} = \sqrt{5}$

$x = 5$

---

2) To complete the square root $\sqrt{60x}$ = an integer

1. Without the x, factor the number to its primes

$\sqrt{6}\ \ \sqrt{10}$

$\sqrt{3}\ \boxed{\sqrt{2}\ \sqrt{2}}\ \sqrt{5}$

2. Figure out what is missing to complete the root

* 2 *

3. The x provides the missing factors

$\sqrt{x} = \sqrt{3} \times \sqrt{5}$

$\sqrt{x} = \sqrt{15}$

$x = 15$

---

3) To complete the cube root $\sqrt[3]{72x}$ = an integer

1. Without the x, factor the number to its primes

$\sqrt[3]{9}\ \ \sqrt[3]{8}$

$\sqrt[3]{3}\ \sqrt[3]{3}\ \boxed{\sqrt[3]{2}\ \sqrt[3]{2}\ \sqrt[3]{2}}$

2. Figure out what is missing to complete the root

* 2

3. The x provides the missing factor

$\sqrt[3]{x} = \sqrt[3]{3}$

$x = 3$

Remember
The cube root has triple factors.

$\sqrt[3]{8} = \sqrt[3]{2 \times 2 \times 2} = 2$

DETAIL — COMPLETING ROOTS

To complete a root, the GMAT phrases this question like this:

"If $2x$ is the square of an integer, ..."

Steps	Example
1. Write down the equation	$2x = (\text{an integer})^2$
2. Take the square root of both sides	$\sqrt{2x} = \text{an integer}$
3. Find the possible values of x:	
x could equal 2	$\sqrt{2x} = \sqrt{4} = 2$
x could equal 8	$\sqrt{2x} = \sqrt{16} = 4$
x could equal 18	$\sqrt{2x} = \sqrt{36} = 6$

The GMAT asks for the value of x in two ways

1. "Which is the lowest possible value of x?"
   Of the possible values of x, {2, 8, 18}, the lowest is 2.

2. "Which of the following fractions must be an integer?"

$$\frac{x}{2} \qquad \frac{x}{3} \qquad \frac{x}{4}$$

With the possible values of x, {2, 8, 18}, only $\frac{x}{2}$ must be an integer.

The other fractions **could** be integers, but only this one **must** be.

QUIZ 1 — COMPLETING ROOTS

If y is the smallest positive integer such that 180 multiplied by y is the square of an integer, then y must be

(A) 2
(B) 5
(C) 6
(D) 7
(E) 9

SOLUTION NEXT PAGE

ARITHMETIC | **QUIZ 1** | **COMPLETING ROOTS**

RADICALS

If y is the smallest positive integer such that 180 multiplied by y is the square of an integer, then y must be

(A) 2
(B) 5
(C) 6
(D) 7
(E) 9

SOLUTION

STEP 1 Separate the information from the question

180 multiplied by y is the square of an integer

STEP 2 Turn this into something you can use

Steps	Example
1. Set up the equation	$180y = (\text{an integer})^2$
2. Take the square root of both sides	$\sqrt{180y}$ = an integer

STEP 3 Use this to answer the question

To complete the square root — $\sqrt{180y}$ = an integer

1. Without the y, factor the number to its primes

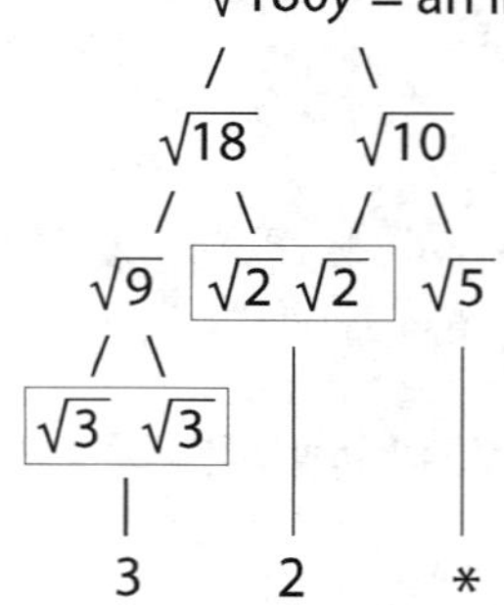

2. Figure out what is missing to complete the root — $\sqrt{y} = \sqrt{5}$

3. The y provides the missing factor — $y = 5$

Answer is B

thinktankSCHOLASTIC

If n and y are positive integers and $60y = n^3$, which of the following must be an integer?

(A) $\dfrac{y}{2^2 \times 3 \times 5}$

(B) $\dfrac{y}{2 \times 3^2 \times 5}$

(C) $\dfrac{y}{2 \times 3 \times 5^2}$

(D) $\dfrac{y}{2^2 \times 3 \times 5^2}$

(E) $\dfrac{y}{2 \times 3^2 \times 5^2}$

SOLUTION NEXT PAGE

ARITHMETIC | **QUIZ 2** | **COMPLETING ROOTS**

RADICALS

If n and y are positive integers and $60y = n^3$, which of the following must be an integer?

(A) $\dfrac{y}{2^2 \times 3 \times 5}$

(B) $\dfrac{y}{2 \times 3^2 \times 5}$

(C) $\dfrac{y}{2 \times 3 \times 5^2}$

(D) $\dfrac{y}{2^2 \times 3 \times 5^2}$

(E) $\dfrac{y}{2 \times 3^2 \times 5^2}$

SOLUTION

STEP 1 Separate the information from the question

$$60y = n^3$$

STEP 2 Turn this into something you can use

Steps	**Example**
1. Take the cube root of both sides	$\sqrt[3]{60y}$ = integer
2. Complete the cube root by factoring the number to its primes	$\sqrt[3]{60y} \to \sqrt[3]{6}$, $\sqrt[3]{10}$; $\sqrt[3]{6} \to \sqrt[3]{3}$, $\sqrt[3]{2}$; $\sqrt[3]{10} \to \sqrt[3]{2}$, $\sqrt[3]{5}$
3. Figure out what factors y provides to complete the cube root	$\sqrt{y} = \sqrt{2} \times \sqrt{3} \times \sqrt{3} \times \sqrt{5} \times \sqrt{5}$ $\sqrt{y} = \sqrt{2 \times 3 \times 3 \times 5 \times 5}$ $y = 2 \times 3^2 \times 5^2$

Remember
The cube root has triple factors:
$\sqrt[3]{8} = \sqrt[3]{2 \times 2 \times 2} = 2$

STEP 3 Use this to answer the question

Substitute $2 \times 3^2 \times 5^2$ for y in each of the answer choices. In only one answer choice will the fraction be an integer.

Remember
A fraction becomes an integer when all the numbers in the bottom cancel out.

Answer is E

TOPIC **ARITHMETIC MEAN** **ARITHMETIC**

STATISTICS

The **average** of a set of numbers is also called the **mean** or **arithmetic mean.** It can be found by adding all the numbers together and dividing by the number of.

LIFELINE 50 **When working with the average, set up the equation with what you have.**

Steps	**Example**
Working with the average	The average of 4, 5, 6, and x is 20
1. Set up the equation	$\frac{4+5+6+x}{4} = 20$
2. Multiply by the denominator	$4+5+6+x = 80$
3. Solve for x	$15+x = 80$
	$x = 65$

QUIZ 1 **ARITHMETIC MEAN**

The average (arithmetic mean) of three numbers is 12. When a fourth number is added, the new average is 10. What number has been added?

(A) 2
(B) 4
(C) 6
(D) 8
(E) 10

SOLUTION NEXT PAGE

ARITHMETIC	QUIZ 1	ARITHMETIC MEAN

STATISTICS

The average (arithmetic mean) of three numbers is 12. When a fourth number is added, the new average is 10. What number has been added?

(A) 2
(B) 4
(C) 6
(D) 8
(E) 10

SOLUTION

STEP 1 Separate the information from the question

The average of three numbers is 12, and when a fourth number is added, the new average is 10.

STEP 2 Turn this into something you can use

Remember
Set up the equation using what you have.

$$\frac{a+b+c}{3} = 12 \quad \text{and} \quad \frac{a+b+c+d}{4} = 10$$

STEP 3 Use this to answer the question

In both equations, multiply both sides by the denominator:

12×3=36   4×10=40

$$a+b+c=36 \quad \text{and} \quad a+b+c+d=40$$

$$d=4$$

Answer is B

If x is the average of the numbers 3, 4, x, 8, and 13, what is the value of x?

(A) 6
(B) 7
(C) 8
(D) 9
(E) 10

SOLUTION NEXT PAGE

ARITHMETIC | STATISTICS

QUIZ 2 **ARITHMETIC MEAN**

If x is the average of the numbers 3, 4, x, 8, and 13, what is the value of x?

(A) 6
(B) 7
(C) 8
(D) 9
(E) 10

SOLUTION

STEP 1 Separate the information from the question

x is the average of 3, 4, x, 8, and 13

STEP 2 Turn this into something you can use

Steps	Example
Set up the equation	$\frac{3+4+x+8+13}{5} = x$

STEP 3 Use this to answer the question

Solve for x

1. Multiply both sides by the denominator	$3+4+x+8+13 = 5x$
2. Subtract x from both sides	$3+4+8+13 = 4x$
3. Add the numbers	$28 = 4x$
4. Divide both sides by 4	$7 = x$

Answer is B

TOPIC	MEDIAN	ARITHMETIC
		STATISTICS

The **median** of a set of numbers is the middle number or the average of the two middle numbers.

LIFELINE 51

When working with the median, put the numbers in order.

Steps	**Example**
Find the median	2, 5, 3, 3, 5, 4, 4
1. Put the numbers in order	2, 3, 3, 4, 4, 5, 5
2. Use the middle number	4
Find the median	5, 2, 3, 4
1. Put the numbers in order	2, 3, 4, 5
2. Average the two middle numbers	3.5

QUIZ MEDIAN

If x and y are positive integers, what is the median of

2, 3, 4, 2, 4, 5, 2, 4, 4, 5, 4, x, y?

(A) 2
(B) 3.5
(C) 4
(D) 4.5
(E) 5

SOLUTION NEXT PAGE

ARITHMETIC	QUIZ	MEDIAN

STATISTICS

If x and y are positive integers, what is the median of

2, 3, 4, 2, 4, 5, 2, 4, 4, 5, 4, x, y?

(A) 2
(B) 3.5
(C) 4
(D) 4.5
(E) 5

SOLUTION

STEP 1 Separate the information from the question

2, 3, 4, 2, 4, 5, 2, 4, 4, 5, 4, x, y

STEP 2 Turn this into something you can use

To find the median,
put the numbers in order 2, 2, 2, 3, 4, 4, 4, 4, 4, 5, 5, x, y (median: the underlined 4)

STEP 3 Use this to answer the question

The median is 4, regardless of what x and y are.

Answer is B

DETAIL	MEDIAN

To find the median, you don't need to know what x and y are.

If x and y are less than 4, the median is 4.

x, y, 2, 2, 2, 3, 4, 4, 4, 4, 4, 5, 5
median: the underlined 4

If x and y are greater than 4, the median is still 4.

2, 2, 2, 3, 4, 4, 4, 4, 4, 5, 5, x, y
median: the underlined 4

If x and y are equal to 4, the median is still 4.

2, 2, 2, 3, 4, x, 4, y, 4, 4, 4, 5, 5
median: the underlined 4

> A common mistake is counting the numbers wrong, such as two 3's instead of one.
>
> If your answer doesn't work, make sure you counted the numbers correctly.

TOPIC	MODE AND RANGE	ARITHMETIC
		STATISTICS

The **mode** is the number that occurs the most often.

The **range** is the distance between the lowest and the highest numbers.

LIFELINE 52

To find the mode, put the numbers in order.

> Mode questions are not common on the GMAT.

Steps	Example
To find the mode of the set	7, 5, 8, 8, 9, 5
1. Put the numbers in order	5, 5, 7, 8, 8, 9
2. Use the most frequently recurring numbers	5 and 8

> This set has two modes. If there are two medians, you average them; if there are two modes, you count them both.

LIFELINE 53

To find the range, subtract the lowest number from the highest number.

Steps	Example
To find the range of the set	14, 11, 32, 27
1. Subtract the lowest number from the highest number	32 − 11
2. The **range** is 21	21

QUIZ — RANGE

If the average of the integers 5, 11, 14, and x is 12, what is the range?

(A) 5
(B) 7
(C) 9
(D) 11
(E) 13

SOLUTION NEXT PAGE

ARITHMETIC | QUIZ | RANGE

STATISTICS

If the average of the integers 5, 11, 14, and x is 12, what is the range?

(A) 5
(B) 7
(C) 9
(D) 11
(E) 13

SOLUTION

STEP 1 Separate the information from the question

The average of the integers 5, 11, 14, and x is 12

STEP 2 Turn this into something you can use

Remember
When working with averages, set up the equation with what you have.

Steps	**Example**
1. Set up the equation	$\frac{5+11+14+x}{4} = 12$
2. Multiply both sides by the denominator:	$5+11+14+x = 48$
3. Add the numbers together	$30+x = 48$
4. Solve for x	$x = 18$

STEP 3 Use this to answer the question

Subtract the lowest number from the the highest number

$$\begin{array}{r} 18 \\ -5 \\ \hline 13 \end{array}$$

The range is 13

Answer is E

TOPIC STANDARD DEVIATION

ARITHMETIC

STATISTICS

STANDARD DEVIATION

Though you don't calculate **standard deviation**, you do have to know how it works.

LIFELINE 54

When working with a standard deviation, start with the mean.

Steps	Example
Working with the standard deviation	The mean is 50 and the standard deviation is 5
1. As two separate equations, add and subtract the standard deviation from the mean	$50 - 5$ $50 + 5$
2. Everything between these two numbers is within the standard deviation	$45 \longleftrightarrow 55$
Working with **two** standard deviations	The mean is 50 and the standard deviation is 5
1. Multiply the standard deviation by 2	$5 \times 2 = 10$
2. As two separate equations, add and subtract this from the mean	$50 - 10$ $50 + 10$
3. Everything between these numbers is within **two** standard deviations	$40 \longleftrightarrow 60$

"Two standard deviations" means "two times the standard deviation."

The same way that "2*x*" means "2 times *x*."

QUIZ STANDARD DEVIATION

The algebra class has an average test score of 83 and a standard deviation of 6. Which of the following test scores is within 1.5 standard deviations from the mean?

(A) 55.5
(B) 60.0
(C) 75.5
(D) 95.0
(E) 99.5

SOLUTION NEXT PAGE

ARITHMETIC

STATISTICS

QUIZ

STANDARD DEVIATION

The algebra class has an average test score of 83 and a standard deviation of 6. Which of the following test scores is within 1.5 standard deviations from the mean?

SOLUTION

STEP 1 Separate the information from the question

The average is 83 and the standard deviation is 6

STEP 2 Turn this into something you can use

Find the lowest and highest values within 1.5 standard deviations

1. Multiply the standard deviation by 1.5 $6 \times 1.5 = 9$

2. As two separate equations, add and subtract this number from mean $\begin{array}{r} 83 \\ -9 \\ \hline \end{array}$ $\begin{array}{r} 83 \\ +9 \\ \hline \end{array}$

3. Everything between these numbers is within 1.5 standard deviations $74 \leftrightarrow 92$

STEP 3 Use this to answer the question

The only answer choice between 74 and 92 is 75.5

Answer is C

DETAIL

STANDARD DEVIATION

The graph on the right is more spread out than the graph on the left. The graph on the right, therefore, has a greater standard deviation.

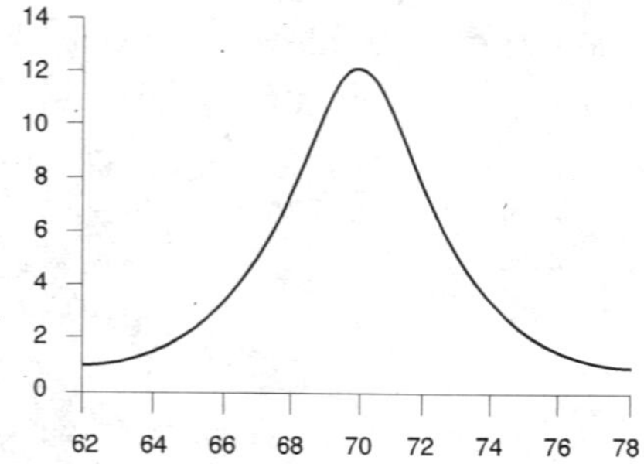

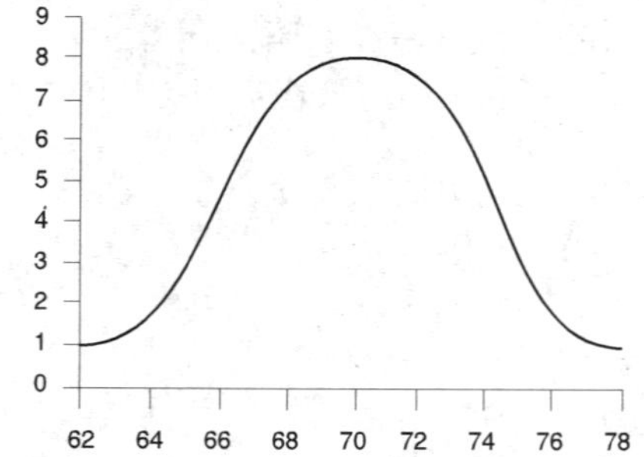

TOPIC	AVERAGE DEVIATION	ARITHMETIC
		STATISTICS

When comparing the standard deviations of two different groups, the **average deviation** is easier to use.

LIFELINE 55

When comparing standard deviations, use the average deviation.

Steps	Example
To find the average deviation of these numbers	5, 7, 10, and 14
1. Average the numbers	$\frac{5+7+10+14}{4} = 9$
2. Find the **positive** distance of each number from the mean	5 7 10 14 / −9 −9 −9 −9 / 4 2 1 5
3. Average these distances:	$\frac{4+2+1+5}{4} = 3$

The average deviation is 3

The Official Guide to the GMAT provides a complex method to calculate the standard deviation.

Use this instead.

QUIZ 1 AVERAGE DEVIATION

This table shows the number of houses sold per real estate agent, per month. For which agent was the standard deviation during these 4 months the greatest?

	Jan.	Feb.	Mar.	Apr.
Agent I	1	5	5	9
Agent II	2	6	6	10
Agent III	8	9	11	12

(A) Agent I
(B) Agent II
(C) Agent III
(D) Agents I and II
(E) Agents II and III

SOLUTION NEXT PAGE

ARITHMETIC

STATISTICS

QUIZ 1 AVERAGE DEVIATION

This table shows the number of houses sold per real estate agent, per month. For which agent was the standard deviation during these 4 months the greatest?

	Jan.	Feb.	Mar.	Apr.
Agent I	1	5	5	9
Agent II	2	6	6	10
Agent III	8	9	11	12

SOLUTION

STEP 1 Separate the information from the question

Agent I: 1, 5, 5, 9

Agent II: 2, 6, 6, 10

Agent III: 8, 9, 11, 12

STEP 2 Turn this into something you can use

Find the average deviation for Agent I:

Steps	Example
1. Average the numbers	$\frac{1+5+5+9}{4} = 5$
2. Find the **positive** distance of each number from the mean	$\begin{array}{cccc} 1 & 5 & 5 & 9 \\ -5 & -5 & -5 & -5 \\ \hline 4 & 0 & 0 & 4 \end{array}$
3. Average the distances:	$\frac{4+0+0+4}{4} = 2$

The average deviation for Agent I is 2. Using the same method, the average deviations for Agents II and III are 2 and 1.5, respectively.

STEP 3 Use this to answer the question

Agents I and II have the highest average deviations and therefore the highest standard deviations.

Answer is D

QUIZ 2 | AVERAGE DEVIATION

If the numbers of gallons of water in 5 different reservoirs have a standard deviation of 10, and 5 gallons of water are added to each reservoir, what is the new standard deviation of the numbers of gallons of water?

(A) 8

(B) 10

(C) 12

(D) 15

(E) 20

SOLUTION NEXT PAGE

ARITHMETIC | **QUIZ 2** | **AVERAGE DEVIATION**

STATISTICS

If the numbers of gallons of water in 5 different reservoirs have a standard deviation of 10, and 5 gallons of water are added to each reservoir, what is the new standard deviation of the numbers of gallons of water?

SOLUTION

STEP 1 Separate the information from the question

The gallons of water in 5 different reservoirs have a standard deviation of 10.

STEP 2 Turn this into something you can use

The average distance from the mean is also 10.

STEP 3 Use this to answer the question

If all the volumes increase by 5 gallons, the average also increases 5 gallons. The distance of each value from the average doesn't change.

Answer is B

DETAIL | **AVERAGE DEVIATION**

You can use the average deviation to prove this. Pick numbers for the gallons in the 5 reservoirs and find the average deviation, then add 5 to each number and find the average deviation again. They will be the same.

Steps	Pick Numbers	5 Added to Each Number
1. Average the numbers	$\frac{5+10+15+20+25}{5} = 15$	$\frac{10+15+20+25+30}{5} = 20$
2. Find the positive distance of each number from the mean	5 10 15 20 25 −15 −15 −15 −15 −15 10 5 0 5 10	10 15 20 25 30 −20 −20 −20 −20 −20 10 5 0 5 10
3. Average the distances:	$\frac{10+5+0+5+10}{5} = 6$	$\frac{10+5+0+5+10}{5} = 6$

The average deviations are the same.

HOMEWORK

ARITHMETIC

13th Editon *GMAT Review Book,* pp. 152-185, question numbers

GMAT REVIEW

ANSWERS NEXT PAGE

Odd × Even
44 C
225 E

Divisibility
5 B
40 A
49 B
74 B
77 C
87 ____
127 ____
174 B
176 ____
204 ____
212 ____
219 ____

Remainders
26 ____
95 ____
118 ____

Units Digit
163 B

Consecutive Integers
90 ____
124 ____
158 ____
172 ____
221 ____

Fractions
2 C
8 A
15 ____
20 ____
27 B
29 ____
32 C
41 B
42 ____
46 B
48 B
76 ____
80 ____
97 ____
108 ____
138 ____
142 ____
142 ____
157 ____
164 ____
187 ____
194 ____
195 ____
209 ____
216 ____
220 ____
226 ____

Decimals
17 ____
85 ____
122 ____
146 ____
199 ____

Percents
6 ____
11 ____
19 ____
21 ____
25 ____
31 ____
57 ____
58 ____
71 ____
84 ____
94 ____
114 ____
115 ____
135 ____
152 ____
171 ____
182 ____
185 ____
189 ____
203 ____

Ratios
22 ____
56 ____
60 ____
82 ____
98 ____
105 ____
113 ____
125 ____
144 ____
151 ____
179 ____
188 ____
200 ____

Exponents
106 ____
111 ____
117 ____
150 ____
170 ____
180 ____
196 ____
217 ____
227 ____
230 ____

Radicals
9 ____
35 ____
52 ____
155 ____

Averages and Medians
1 ____
12 ____
16 ____
30 ____
53 ____
67 ____
101 ____
109 ____
132 ____
145 ____
183 ____

Standard Deviation
112 ____

225: 3(3+1)(3+2)
3(4)·3(5)=188
188÷4=47

ARITHMETIC	HOMEWORK
GMAT REVIEW	**13th Editon** *GMAT Review Book,* pp. 152-185, question answers
ANSWERS	

Odd × Even
44 ____C
225 ____E

Divisibility
5 ____B
40 ____A
49 ____C
74 ____B
77 ____C
87 ____A
127 ____B
174 ____D
176 ____E
204 ____C
212 ____A
219 ____D

Remainders
26 ____B
95 ____B
118 ____E

Units Digit
163 ____A

Consecutive Integers
90 ____A
124 ____B
158 ____C
172 ____B
221 ____E

Fractions
2 ____C
8 ____D
15 ____C
20 ____C
27 ____D
29 ____B
32 ____C
41 ____A
42 ____B
46 ____B
48 ____B
76 ____E
80 ____A
97 ____E
108 ____D
138 ____E
141 ____A
142 ____D
157 ____C
164 ____A
187 ____E
194 ____D
195 ____C
209 ____A
216 ____C
220 ____D
226 ____D

Decimals
17 ____B
85 ____A
122 ____B
146 ____E
199 ____D

Percents
6 ____E
11 ____C
19 ____D
21 ____E
25 ____A
31 ____B
57 ____B
58 ____D
71 ____D
84 ____C
94 ____C
114 ____B
115 ____C
135 ____B
152 ____B
171 ____D
182 ____D
185 ____D
189 ____A
203 ____C

Ratios
22 ____A
56 ____D
60 ____D
82 ____D
98 ____D
105 ____E
113 ____B
125 ____C
144 ____E
151 ____E
179 ____B
188 ____C
200 ____E

Exponents
106 ____B
111 ____B
117 ____C
150 ____E
170 ____B
180 ____D
196 ____E
217 ____B
227 ____A
230 ____C

Radicals
9 ____E
35 ____B
52 ____A
155 ____E

Averages and Medians
1 ____A
12 ____B
16 ____A
30 ____E
53 ____A
67 ____B
101 ____B
109 ____C
132 ____B
145 ____D
183 ____D

Standard Deviation
112 ____D

Algebra

Algebra

Lesson 2

Algebra is based on the concept that x can have more than one value.

For example, if $x^2 = 16$, $x = 4$ or -4.
x has one value: You don't know which it is, but it's one of these two.

The idea of x having more than one value, or a changing value, opens a wide range of possibilities.

TOPIC **ABSOLUTE VALUE** | ALGEBRA

NUMBER LINE

That *x* has more than one value begins on the number line.

Absolute value refers to the **distance from zero** on the number line.

The absolute value of 3.5 literally means it is 3.5 from zero.

This is why the absolute value of a negative number has a positive value. Since –3.5 is also 3.5 away from zero, the absolute value of –3.5 is 3.5.

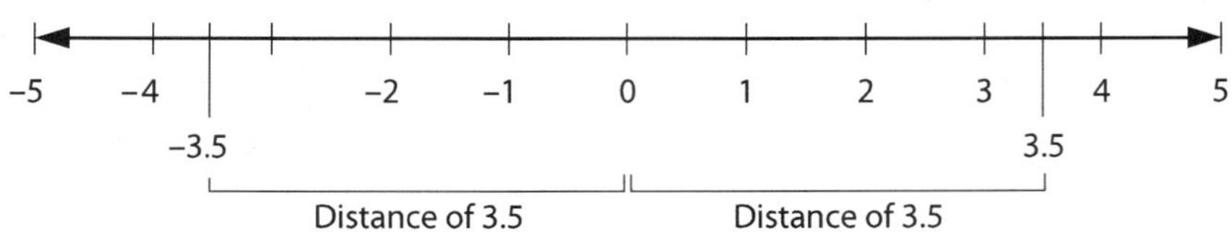

This is how the number line gives *x* more than one value.
If $|x| = 4$, then *x* is 4 away from zero, and it could equal 4 or – 4.

Remember that *x* has one value, and you don't know which it is.

LIFELINE **56**

When working with the number line, draw the number line.

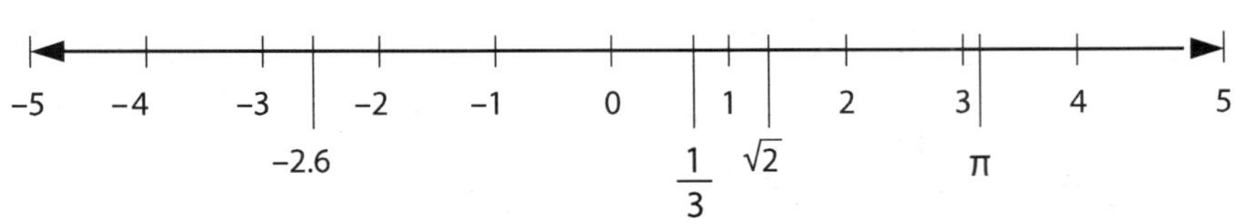

The number line contains every real number, including repeating decimals such as *pi* or 1/3.

ALGEBRA

NUMBER LINE

LIFELINE 57

When x is inside the absolute value, place it on both sides of the number line.

If $|x - 3| = 4$, what could x be?

There are two ways to solve:

First method:

$|x - 3|$ means $x - 3$ is 4 away from zero.

1. Draw the number line

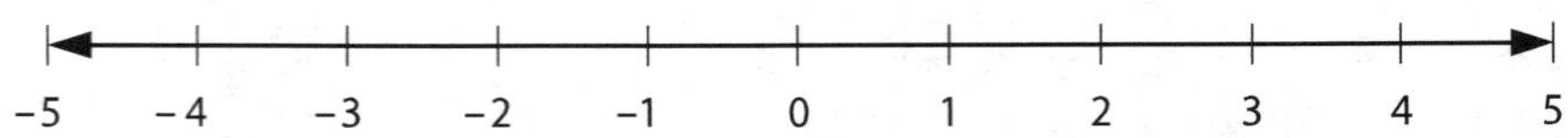

2. Write $x - 3$ (without the absolute value bars) at 4 and −4

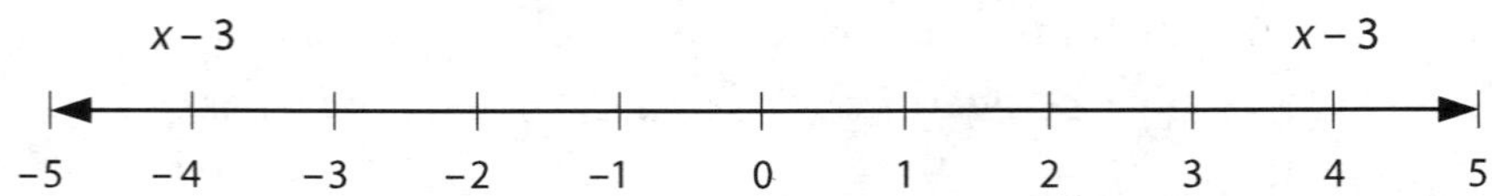

3. Solve for x using the two different equations.

$$\begin{array}{rcl} x - 3 &=& -4 \\ +3 && +3 \\ \hline x &=& -1 \end{array} \qquad\qquad \begin{array}{rcl} x - 3 &=& 4 \\ +3 && +3 \\ \hline x &=& 7 \end{array}$$

Second method:

1. Copy down the equation $\quad |x - 3| = 4$

2. Drop the absolute value bars and set $x - 3$ equal to both 4 and − 4

3. Add 3 to all sides

$$\begin{array}{rcl} x - 3 &=& 4,\ -4 \\ +3 && +3\ \ +3 \\ \hline x &=& 7,\ -1 \end{array}$$

DRILL **ABSOLUTE VALUE**

If $|x + 3| = 9$, what could x be?

SOLUTION NEXT PAGE

DRILL	ABSOLUTE VALUE	ALGEBRA

If $|x + 3| = 9$, then what could x be?

Answer:

$$\begin{array}{rcl} x + 3 & = & 9,\ -9 \\ -3 & & -3\ \ -3 \\ \hline x & = & 6,\ -12 \end{array}$$

QUIZ	ABSOLUTE VALUE

If $|x^2 - 5| = 4$, then the sum of the absolute values of all possible values of x is

(A) 1
(B) 2
(C) 4
(D) 6
(E) 8

NUMBER LINE

SOLUTION NEXT PAGE

ALGEBRA | **QUIZ** | **ABSOLUTE VALUE**

NUMBER LINE

If $|x^2 - 5| = 4$, then the sum of the absolute values of the possible values of x is

(A) 1
(B) 2
(C) 4
(D) 6
(E) 8

SOLUTION

STEP 1 | **Separate the information from the question**

$|x^2 - 5| = 4$

STEP 2 | **Turn this into something you can use**

Steps	Example		
1. Write out the two equations without the absolute value bars.	$x^2 - 5 = 4$	and	$x^2 - 5 = -4$
2. Add 5 to all the sides:	$x^2 - 5 = 4$ $+5 \quad +5$ $x^2 = 9$	and	$x^2 - 5 = -4$ $+5 \quad +5$ $x^2 = 1$
3. Square root both sides.	$x = 3, -3$		$x = 1, -1$

Remember
A square root has two answers: one positive and one negative.

STEP 3 | **Use this to answer the question**

The absolute values of these are 3, 3, 1, 1

The sum is 8

Answer is E

DETAIL | **ABSOLUTE VALUE**

With math inside absolute value bars, treat the bars like parentheses, and do the math first.

For example: $|5 - 7|$ becomes $|-2|$

Also, $|x|$ is equivalent to $\sqrt{(x^2)}$

For example, if $x = -5$, then $|x| = 5$, and $\sqrt{(x^2)} = 5$

TOPIC **INEQUALITIES** — **ALGEBRA**

NUMBER LINE

The **inequality** is also a number line concept.

For example, if $x > 5$, then x has one value greater than 5.

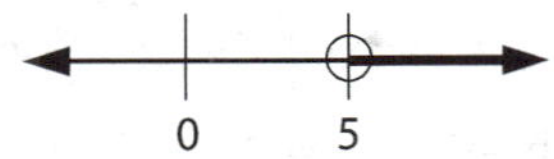

The inequality also makes a statement.
For example, $x > 0$ means x is positive.

- $>$ Greater Than
- $\geq$ Greater Than or Equal To
- $<$ Less Than
- $\leq$ Less Than or Equal To
- $\neq$ Not Equal To

LIFELINE 58

When working with an inequality, for the most part treat it like an equal sign.

When dividing both sides by a **positive** number, the inequality stays the same.

$$\frac{3x}{3} > \frac{15}{3}$$
$$x > 5$$

When dividing both sides by a **negative** number, switch the inequality.

$$\frac{-5x}{-5} > \frac{20}{-5}$$
$$x < -4$$

When dividing both sides by an **unknown**, and you don't know whether it's positive or negative, you don't know whether to switch the sign, so you can't solve it.

$$\frac{xy}{y} > \frac{5y}{y}$$
$$x \ ? \ 5$$

When taking the reciprocal of fractions, switch the inequality.

$$\frac{3}{8} > \frac{2}{8}$$
$$\frac{8}{3} < \frac{8}{2}$$

ALGEBRA | **DETAIL** | **INEQUALITIES**

NUMBER LINE

The negative does not always switch the inequality.

When **multiplying** or **dividing** by a negative, switch the inequality.

$$\begin{array}{rcr} 3 & < & 5 \\ \times\ -2 & & \times\ -2 \\ \hline -6 & > & -10 \end{array}$$

When **adding** or **subtracting** a negative, keep the inequality the same.

$$\begin{array}{rcr} 7 & < & 10 \\ -\ 2 & & -\ 2 \\ \hline 5 & < & 8 \end{array}$$

Taking the **square root** of an inequality produces two answers.

The way that a square root produces positive and negative answers ...

$$x^2 = 16 \quad \rightarrow \quad x = 4 \quad \text{or} \quad x = -4$$

switch the inequality for the negative answer.

$$x^2 > 25 \quad \rightarrow \quad x > 5 \quad \text{or} \quad x < -5$$

DRILL | **INEQUALITIES**

Solve for x:

1) $-5x > -25$

2) $x + 12 < 22$

3) $x^2 < 49$

SOLUTION NEXT PAGE

DRILL — INEQUALITIES — Solutions

ALGEBRA

NUMBER LINE

1) $-5x > -25$

Divide both sides by -5

and switch the inequality.

$$\frac{5x}{-5} > \frac{-25}{-5}$$

$$x < 5$$

2) $x + 12 < 22$

Subtract 12 from both sides

and keep the inequality the same

$$x + 12 < 22$$
$$-12 \quad -12$$
$$x < 10$$

3) $x^2 < 49$

Square root both sides

and switch the inequality for the negative answer.

$$x^2 < 49$$

$$\sqrt{x^2} < \sqrt{49}$$

$$x < 7 \text{ or } x > -7$$

QUIZ — INEQUALITIES

Given the inequality $\dfrac{-6 + 3x}{-2} < -3x + 7.5$

which of the following could be the value of x?

(A) 1 only
(B) 1 and 2
(C) 2 and 3
(D) 3 only
(E) neither 1, 2, nor 3

SOLUTION NEXT PAGE

ALGEBRA | **QUIZ** | **INEQUALITIES**

NUMBER LINE

Given the inequality $\dfrac{-6+3x}{-2} < -3x + 7.5$

which of the following could be the value of x?

(A) 1 only
(B) 1 and 2
(C) 2 and 3
(D) 3 only
(E) neither 1, 2, nor 3

SOLUTION

STEP 1 | **Separate the information from the question**

Copy down the equation $\dfrac{-6+3x}{-2} < -3x + 7.5$

STEP 2 | **Turn this into something you can use**

Steps

1. Multipy both sides by –2 $\quad -6 + 3x > 6x - 15$

   a. cancel the fraction on the left
   b. double and switch the signs of the quantity on the right
   c. switch the inequality

2. Move the numbers to the right by adding 6 to both sides.

$$\begin{array}{rcl} -6+3x & > & 6x - 15 \\ +6 & & +6 \\ \hline 3x & > & 6x - 9 \end{array}$$

3. Move the x's to the left by subtracting $6x$ from both sides.

$$\begin{array}{rcl} 3x & > & 6x - 9 \\ -6x & > & -6x \\ \hline -3x & > & -9 \end{array}$$

4. Divide both sides by –3 and switch the inequality again

$$\frac{-3x}{-3} > \frac{-9}{-3}$$

$$x < 3$$

STEP 3 | **Use this to answer the question**

If x is less than 3, then of the choices (1, 2, and 3), it could only be 1 or 2.

Answer is B

COORDINATE GEOMETRY

Coordinate Geometry refers to two number lines that are perpendicular. This is also called the **coordinate grid** or the **rectangular coordinate system.**

The grid is divided into four quadrants, or regions, titled I, II, III, and IV.

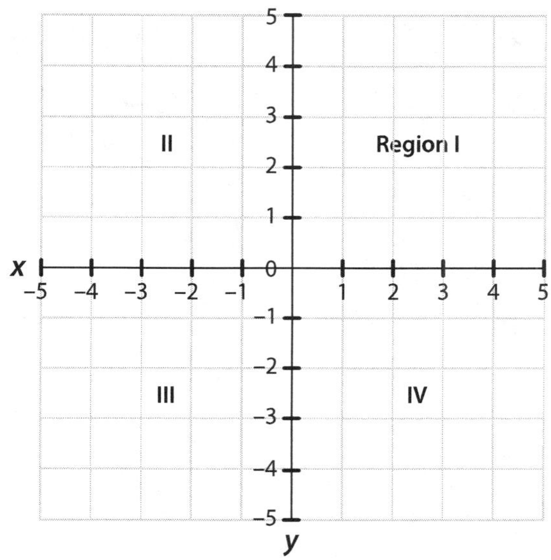

The center of the grid, usually marked with a zero, is called the **origin**.

Don't worry about memorizing the regions. When necessary, the GMAT provides a drawing of which is which.

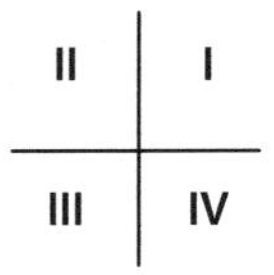

Below, the (*x*, *y*) coordinates of points *A* and *B* are (2, 1), and (3, –2), respectively.

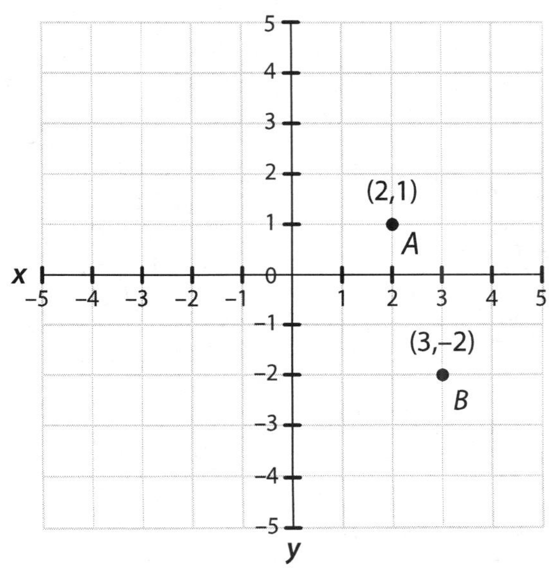

ALGEBRA

COORDINATE GEOMETRY

LIFELINE 59

When drawing the line from an equation, pick a value for *x*, and find the corresponding value of *y*.

You won't be drawing many lines from equations: you'll be finding more equations from lines. But this helps you grasp how it works.

Steps	Example
To draw the line for the equation	$y = 2x + 1$

1. Draw an *x/y* table

x	y

2. Put down three numbers for *x*

x	y
0	
1	
2	

3. Find the corresponding values of *y*

x	y
0	1
1	3
2	5

4. Plot these three (x,y) points on the grid.

5. Draw the line.

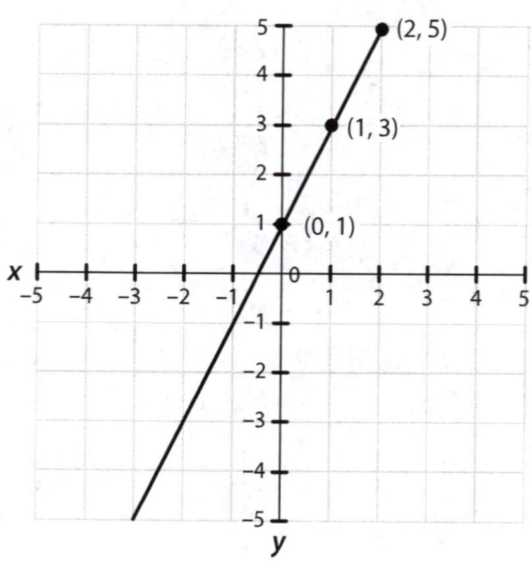

The unknowns don't have to be *x* and *y*.

For example, $b = 2a + 1$ is also considered a linear equation.

TOPIC FINDING THE EQUATION FROM THE LINE **ALGEBRA**

COORDINATE GEOMETRY

The two things that define a line are where it is and which way it is pointing.

What it is:	**What it's called:**
1. Where the line is	***y*-intercept,** or "*b*"
2. Which way it's pointing	**the slope,** or "*m*"

These two things are in the linear equation:

The *y* and *x* change, but the *m* and *b* are constant.

$$y = mx + b$$

(*mx*: slope; *b*: *y*-intercept)

***y*–intercept** refers to where the line crosses the *y* axis.

Constant means the value doesn't change.

LIFELINE 60

To find the equation of a line, find the *y*-intercept and the slope.

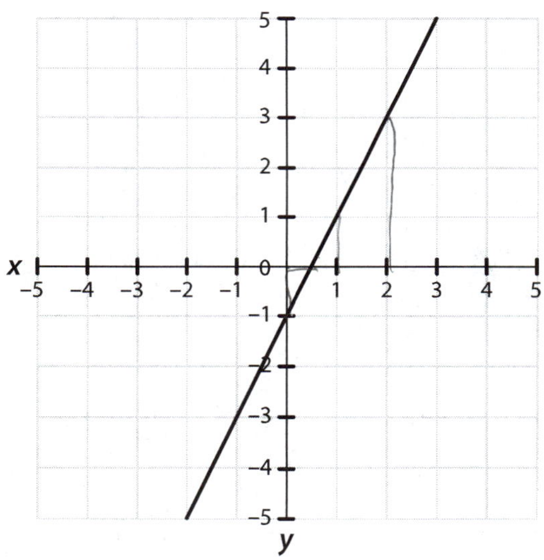

Steps	**Example**
1. Write down the equation	$y =$ ___ $x +$ ___
2. Find the *y*-intercept and write that for the "*b*" in the equation	$y =$ ___ $x +$ -1
3. Find the slope (next page) and write that for the "*m*" in the equation	$y = 2x + -1$

And you have the equation for the line. $y = 2x - 1$

The linear equation $y = 2x - 1$ tells us that the line has a **slope** of 2 and a ***y*-intercept** of –1.

ALGEBRA

COORDINATE GEOMETRY

LIFELINE 61

When finding the slope of a line, use "rise over run."

Don't use the slope formula: $\dfrac{y_2 - y_1}{x_2 - x_1}$

This works, but here's an easier way.

Step 1
Draw a triangle **underneath** the line.
It can be anywhere, as long as ...

a) It touches the line **where the line crosses the grid**

b) It's **underneath** the line

Because 6/3 equals 2/1, both of these triangles will work.

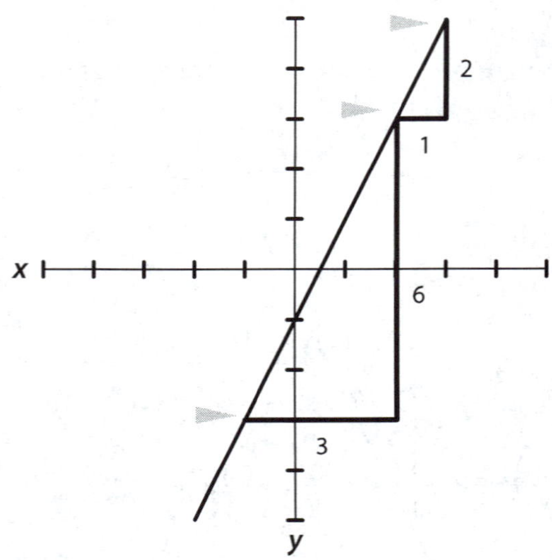

Step 2
Count the spaces on the grid for the sides of the triangle

Step 3
Using "rise over run", place these numbers in a fraction

$\dfrac{2}{1}$ (smaller triangle) or $\dfrac{6}{3}$ (larger triangle)

Step 4
Determine whether the slope is positive or negative.

Going left to right:

If the line goes up, the slope is positive

If the line goes down, the slope is negative

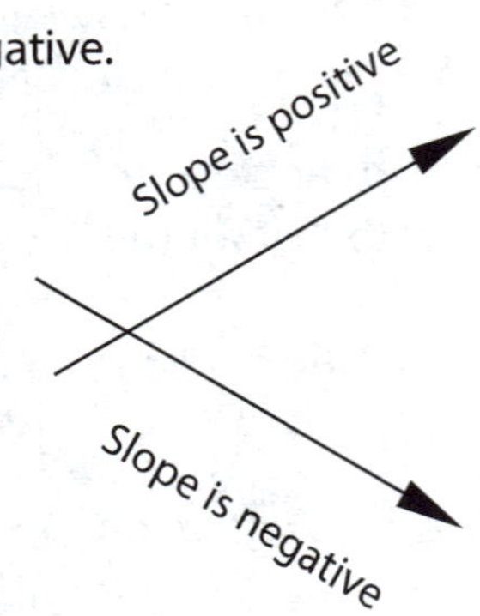

This line has a positive slope: 2

DRILL | FINDING THE EQUATION FROM THE LINE | ALGEBRA

COORDINATE GEOMETRY

Find the equation for this line.

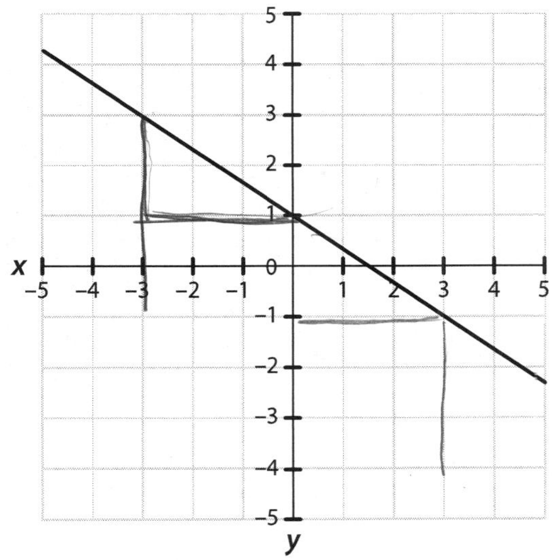

$y = -\frac{2}{3}x + 1$

SOLUTION NEXT PAGE

ALGEBRA	DRILL	FINDING THE EQUATION FROM THE LINE

COORDINATE GEOMETRY

SOLUTION

Find the slope.

1. Draw a triangle underneath the line

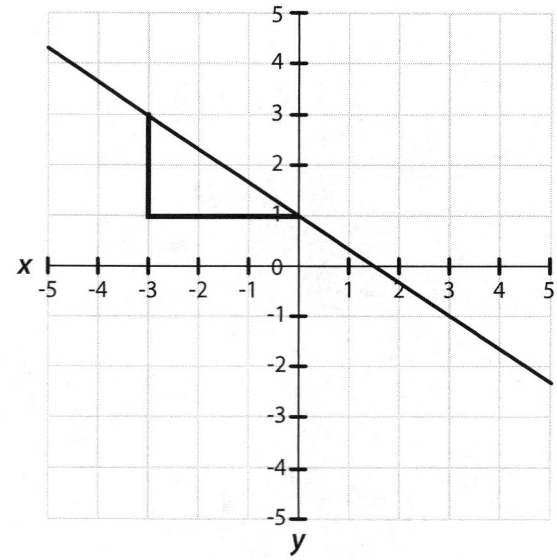

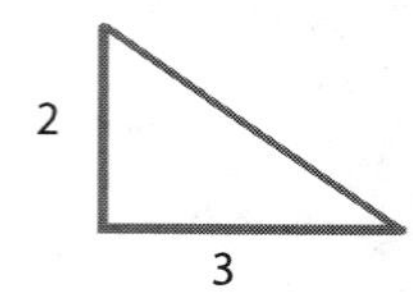

2. Count the spaces on the grid for the sides of the triangle: 2 and 3
3. Using "rise over run", place these numbers in a fraction: $\frac{2}{3}$
4. Determine whether the slope is positive or negative.

   The line is going down, so the slope is negative: $-\frac{2}{3}$

Find the equation:

1. Write down the equation $y =$ ______ $x +$ ______
2. Find the y-intercept and write that for the "b" in the equation $y =$ ______ $x +$ $\underline{\ 1\ }$
3. Find the slope and write that for the "m" in the equation $y = \underline{-\frac{2}{3}}\ x + \underline{\ 1\ }$

And the formula for this line is $y = -\frac{2}{3}x + 1$

 thinktankSCHOLASTIC

TOPIC | CONVERTING AN EQUATION TO $y = mx + b$ | **ALGEBRA**

COORDINATE GEOMETRY

Any equation with two unknowns (and no exponents) is a linear equation. Before working with it, convert it to the form of $y = mx + b$.

LIFELINE 62

When converting a linear equation to the form "$y = mx + b$", basically set y equal to everything else.

For example, set the equation as y = (everything else)

$$2x + 3y - 3 = 0 \quad \rightarrow \quad y = -\frac{2}{3}x + 1$$

This is called the **slope-intercept** form because it highlights the slope and intercept.

QUIZ

CONVERTING AN EQUATION TO $y = mx + b$

In the coordinate system below, which of the following is the equation of the line?

(A) $3y - x + 3 = 0$

(B) $3y - x - 3 = 0$

(C) $3y + x + 3 = 0$

(D) $y + 3x + 3 = 0$

(E) $y - 3x - 3 = 0$

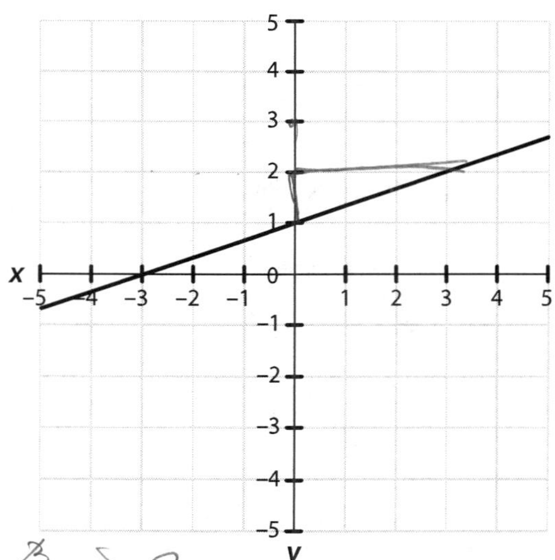

SOLUTION NEXT PAGE

ALGEBRA | QUIZ | CONVERTING AN EQUATION TO $y = mx + b$

COORDINATE GEOMETRY

In the coordinate system below, which of the following is the equation of the line?

(A) $3y - x + 3 = 0$

(B) $3y - x - 3 = 0$

(C) $3y + x + 3 = 0$

(D) $y + 3x + 3 = 0$

(E) $y - 3x - 3 = 0$

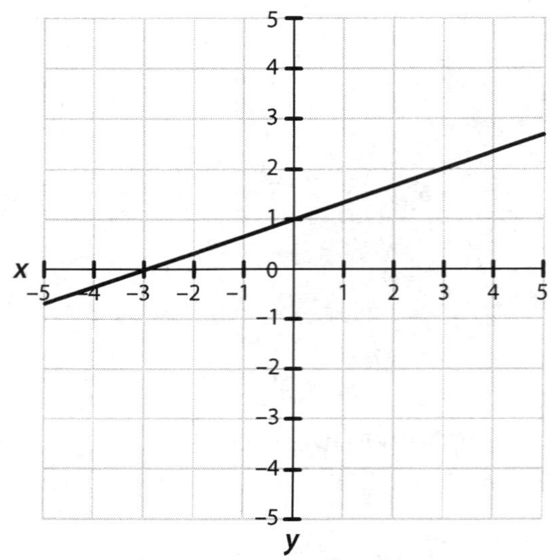

SOLUTION

STEP 1 **Separate the information from the question**

Copy the drawing to your scratch paper.

STEP 2 **Turn this into something you can use**

Find the equation of the line:

1. Write down the equation $y =$ ______ $x +$ ______

2. Find the y-intercept and write that for the "b" in the equation $y =$ ______ $x +$ 1

3. Find the slope and write that for the for the "m" in the equation $y = \frac{1}{3} x + 1$

And the equation for this line is $y = \frac{1}{3} x + 1$

STEP 3 **Use this to answer the question**

Convert the equation to match the answer choices.

1. Multiply both sides by 3 to get rid of the fraction: $3y = x + 3$
2. Move everything to the left of the equal sign $3y - x - 3 = 0$

Answer is B

TOPIC PARABOLAS

ALGEBRA

COORDINATE GEOMETRY

An equation with two unknowns creates a line.

$$y = 2x + 1$$

Adding an exponent curves the line, and in this case, creates a **parabola**.

$$y = x^2$$

A "line" is straight, so when the drawing curves, it's no longer a "line."

Drawing a parabola $y = x^2$

Steps / **Example**

1. Draw an *x/y* table

x	y

2. Put down positive and negative numbers for *x*

x	y
-2	
-1	
0	
1	
2	

3. Find the corresponding values of *y*

x	y
-2	4
-1	1
0	0
1	1
2	4

4. By plotting these (*x*,*y*) points on the grid, you have a parabola:

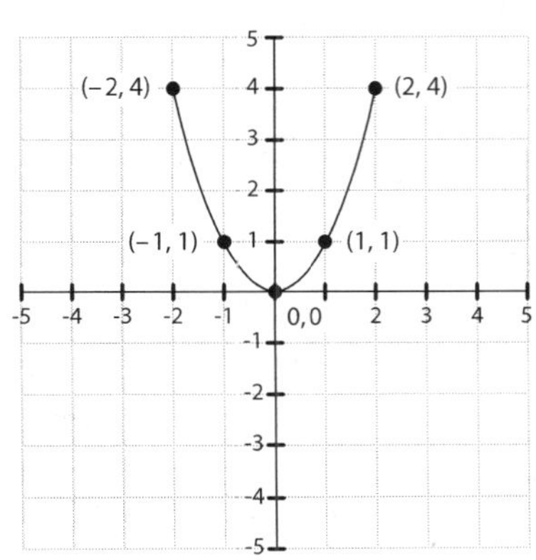

CONTINUED NEXT PAGE

ALGEBRA | DETAIL | PARABOLAS

COORDINATE GEOMETRY

There are two things you need to know about the parabola:

1) It comes from an equation with an exponent.

2) It's symmetrical.

LIFELINE 63

To solve a parabola, find the x and y values of one point, and plug them into the equation.

Parabola questions are rare on the GMAT and typically feature a parabola pointing up, like a "lucky" horseshoe U not down ∩ or sideways ⊂

Think of it as a "lucky" parabola.

The parabola $y = x^2$ can be varied by adjusting the formula. For example:

$y = x^2 + 2$ moves the parabola up 2 spaces.

$y = 2x^2$ makes the parabola narrower by half.

$y = -x^2$ turns the parabola upside down.

However, knowing these variations won't help on the GMAT.

Instead, approach these questions like puzzles, by filling in the pieces you know and solving for what you don't know.

You can always find the x and y values of at least one point for plugging into the equation.

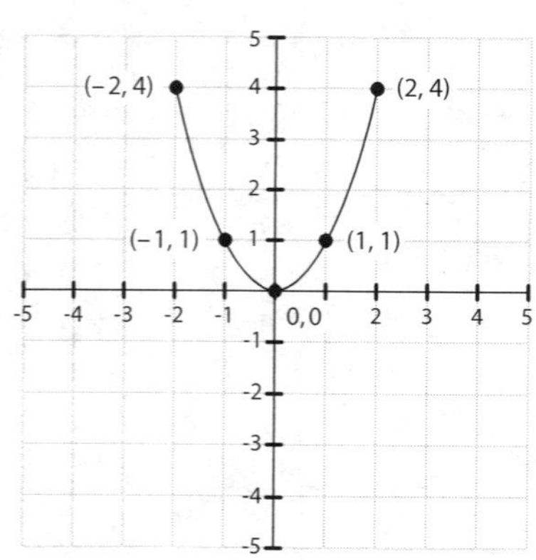

COORDINATE GEOMETRY

In the figure below, *ABCD* is a square, and points *A*, *B*, and *O* lie on the graph of $y = kx^2$, where k is a constant. If the area of the square is 100, what is the value of k ?

(A) $\frac{1}{5}$

(B) $\frac{2}{5}$

(C) $\frac{3}{5}$

(D) $\frac{4}{5}$

(E) 1

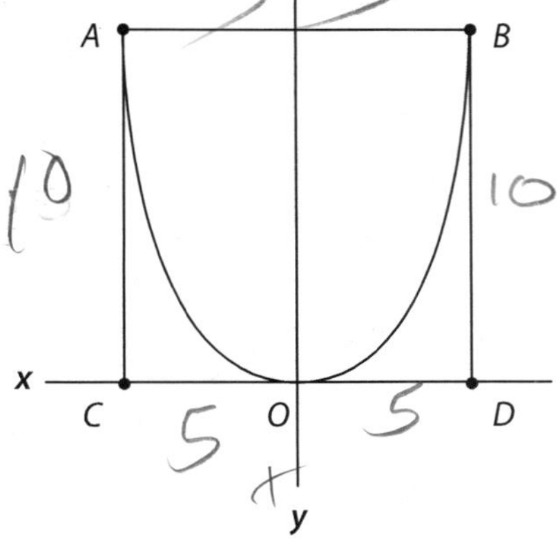

SOLUTION NEXT PAGE

ALGEBRA | **QUIZ** | **PARABOLAS**

COORDINATE GEOMETRY

In the figure below, ABCD is a square, and points A, B, and O lie on the graph of $y = kx^2$, where k is a constant. If the area of the square is 100, what is the value of k ?

(A) $\frac{1}{5}$

(B) $\frac{2}{5}$

(C) $\frac{3}{5}$

(D) $\frac{4}{5}$

(E) 1

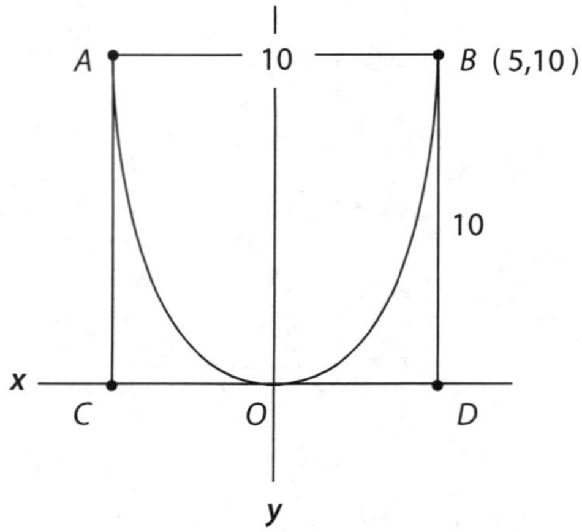

Remember
Find the x and y values of one point and plug them into the equation.

SOLUTION

STEP 1 Separate the information from the question

The graph is $y = kx^2$ and the area of the square is 100

STEP 2 Turn this into something you can use

Each side of the square is 10, making the (x, y) coordinates of point B (5, 10).

STEP 3 Use this to answer the question

For point B, when $x = 5, y = 10$.

Plug these into the equation: $y = kx^2$

$10 = k(5)^2$

And solve for k $10 = k25$

$\frac{10}{25} = k$

$\frac{2}{5} = k$

Answer is B

With one linear equation, there are infinite values of x and y.
For each value of x, there is a corresponding value of y.

$$y = 2x + 1$$

With two linear equations, you can usually, but not always, solve for x and y. There are three possibilities.

Lines Cross:	Same Line:	Parallel Lines:
$x + y = 3$	$x + y = 3$	$x + y = 3$
$2x + y = 5$	$2x + 2y = 6$	$x + y = 5$
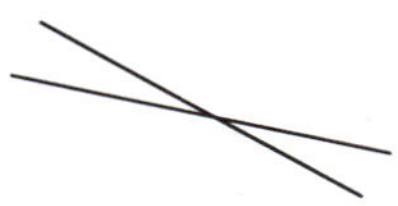		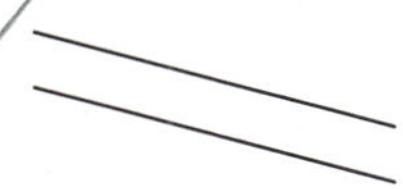
Able to Solve	Can't Solve	Can't Solve

The lines cross if they have different slopes. Most GMAT questions on this use lines that cross.

LIFELINE 64

You need two equations to solve for two unknowns.

Having two equations and two unknowns is called **simultaneous equations.** There are two ways to solve these, and the one that you use depends on the arrangement of the equations.

Method 1: Substitution

Steps	Example
1. Arrange at least one of the equations so that y is by itself	$y = 2x + 5$ $y = x + 3$
2. Because y equals $2x + 5$, substitute $2x + 5$ for y in the second equation	$2x + 5 = x + 3$
3. Solve for x	$x = -2$
4. Plug the value of x into the original equations. You should get the same value of y in both equations.	$y = 2(-2) + 5 \rightarrow y = 1$ $y = (-2) + 3 \rightarrow y = 1$

LIFELINE CONTINUED

ALGEBRA | **LIFELINE** | **SIMULTANEOUS EQUATIONS — Continued**

COORDINATE GEOMETRY

When the x's, y's, numbers, and equal sign are lined up, use the subtraction method. Otherwise use the substitution method.

Method 2: Subtraction

Steps	Example
1. Arrange the equations so that the x's, y's, numbers, and equal signs line up.	$y = 2x + 5$ $y = x + 3$
2. Subtract one entire equation from the other, piece by piece.	$y = 2x + 5$ $-(y = x + 3)$ $0 = x + 2$
3. Solve for x	$x = -2$
4. Plug the value of x into the original equations. You should get the same value of y in both equations.	$y = 2(-2) + 5$ $y = 1$ $y = (-2) + 3$ $y = 1$

Don't distribute the negative in the second equation. Just put brackets and a minus sign.

DETAIL | **SIMULTANEOUS EQUATIONS**

If you draw the lines from $y = 2x + 5$ and $y = x + 3$, the (x, y) coordinates of where they cross are $(-2, 1)$.

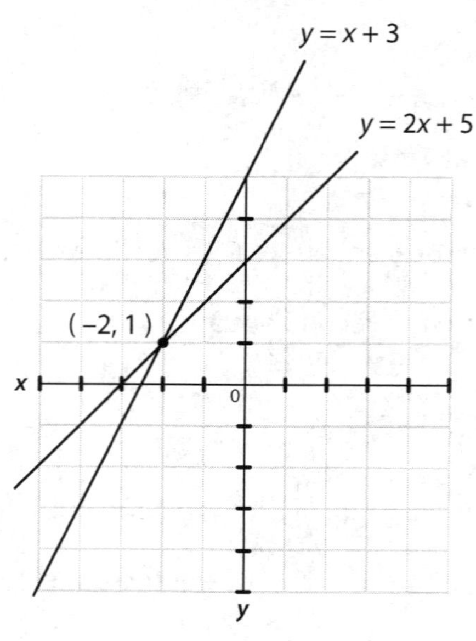

1) Solve for x and y by substituting the x

$$y = 2x + 1$$
$$y = x + 5$$

$y = 2x + 1$
$- \; y = x + 5$
$x = 4$

$y = 8 + 1$
$y = 9$

2) Solve for x and y by subtracting one of the equations.

$$y = 3x + 2$$
$$y = 2x + 5$$

$x = 3$

$y = 9 + 2 = 11$

$y = 11$

SOLUTION NEXT PAGE

ALGEBRA | DRILL | SIMULTANEOUS EQUATIONS

COORDINATE GEOMETRY

1) Solving for x and y by substituting the x

$y = 2x + 1$
$y = x + 5$

SOLUTION

Steps	Example
1. Arrange at least one of the equations	$y = 2x + 1$
so that y is by itself	$y = x + 5$
2. Because y equals $2x + 1$, substitute	
$2x + 1$ for y in the second equation	$2x + 1 = x + 5$
3. Solve for x	$x = 4$
4. Plug the value of x into the original equations. You get the same	$y = 2(4) + 1$ and $y = (4) + 5$
value of y in both equations.	$y = 9$

2) Solve for x and y by subtracting one of the equations.

$y = 3x + 2$
$y = 2x + 5$

SOLUTION

Steps	Example
1. Arrange the equations so that the	$y = 3x + 2$
x's, y's, numbers, and equal signs	$y = 2x + 5$
are lined up	
2. Subtract one entire equation	$y = 3x + 2$
from the other, piece by piece.	$-(y = 2x + 5)$
	$0 = x - 3$
3. Solve for x	$x = 3$
4. Plug the value of x into the original	$y = 3(3) + 2$ and $y = 2(3) + 5$
equations. You get the same	$y = 11$
value of y in both equations.	

If Kim had half the stamps in her collection, she would have 20 fewer stamps than John. If John has 20 fewer stamps than Kim, how many stamps does Kim have?

(A) 20
(B) 40
(C) 60
(D) 80
(E) 100

SOLUTION NEXT PAGE

ALGEBRA	QUIZ	SIMULTANEOUS EQUATIONS

COORDINATE GEOMETRY

If Kim had half the stamps in her collection, she would have 20 fewer stamps than John. If John has 20 fewer stamps than Kim, how many stamps does Kim have?

(A) 20
(B) 40
(C) 60
(D) 80
(E) 100

SOLUTION

STEP 1 **Separate the information from the question**

If Kim had half the stamps in her collection,

she would have 20 fewer stamps than John.

John has 20 fewer stamps than Kim.

STEP 2 **Turn this into something you can use**

$$\tfrac{1}{2}k = j - 20$$
$$j = k - 20$$

STEP 3 **Use this to answer the question**

Because $j = k - 20$, substitute $k - 20$ for j in the first equation.

$$\tfrac{1}{2}k = (k - 20) - 20$$

$$\tfrac{1}{2}k = k - 40$$
$$-k \qquad -k$$
$$-\tfrac{1}{2}k = -40$$

$$k = 80$$

Answer is D

TOPIC	MULTIPLYING EXPRESSIONS	ALGEBRA

EXPRESSIONS

Expressions aren't multiplied as often they are factored, but knowing how to multiply them helps understand how to factor them.

LIFELINE 65

When multiplying expressions, multiply everything in one bracket by everything in the other bracket.

Steps	Example
1. Write out the two expressions	$(x + 4)(x - 2) = 0$
2. Multiply the first term (x) by everything in the second expression (x and -2)	$(x + 4)(x - 2)$ $x^2 - 2x$
3. Multiply the second term (4) by everything in the second expression (x and -2)	$(x + 4)(x - 2)$ $4x - 8$
4. Combine your work into a final equation.	$x^2 + 2x - 8 = 0$

This is also known as the FOIL method: First, Out, In, Last

DRILL — MULTIPLYING EXPRESSIONS

Multiply these expressions: $(2x + 3)(x - 4) = 0$

QUIZ — MULTIPLYING EXPRESSIONS

If $x^2 = 10$, $(x + 5)(x - 5) =$

(A) -5
(B) -10
(C) -15
(D) -20
(E) -25

SOLUTION NEXT PAGE

ALGEBRA

EXPRESSIONS

DRILL — MULTIPLYING EXPRESSIONS

Multiply these expressions using the FOIL method

$(2x + 3)(x - 4) = 0$

$$\begin{array}{r} 2x^2 - 8x \\ 3x - 12 \\ \hline 2x^2 - 5x - 12 = 0 \end{array}$$

Write these so that the like terms (the *x*'s) are stacked, making them easier to combine.

QUIZ — MULTIPLYING EXPRESSIONS

If $x^2 = 10$, $(x + 5)(x - 5) =$

(A) – 5
(B) – 10
(C) – 15
(D) – 20
(E) – 25

SOLUTION

STEP 1 Separate the information from the question

$(x + 5)(x - 5)$

STEP 2 Turn this into something you can use

Foil this into
$$\begin{array}{r} x^2 - 5x \\ + 5x - 25 \\ \hline x^2 - 25 \end{array}$$

STEP 3 Use this to answer the question

$x^2 = 10$, so this becomes $10 - 25 \rightarrow -15$

Answer is C

TOPIC	FACTORING EXPRESSIONS	ALGEBRA
		EXPRESSIONS

Factoring expressions is the opposite of multiplying them.

LIFELINE 66

With x^2, find the two possible values of x.

To factor the expression $x^2 - x = 6$

Steps	**Example**
1. Set it equal to zero	$x^2 - x - 6 = 0$
2. Put two sets of parentheses underneath the equation	$(\quad)(\quad)$
a. Because $(x)(x) = x^2$, place the x's	$(x \quad)(x \quad)$
b. Because $(2)(3) = 6$, put 2 and 3 in the parentheses, on the right.	$(x \quad 2)(x \quad 3)$
c. Determine whether the numbers are positive or negative.	$(x + 2)(x - 3)$

3. To find the values of x that make this equal zero ...

a. Cover the first expression, and $x = 3$	$\boxed{}(x - 3) = 0$
b. Cover the second expression, and $x = -2$	$(x + 2)\boxed{} = 0$

Whether x is 3 or -2, the equation equals zero, and these two possible values of x are the **roots** of the equation.

x doesn't equal **both** 3 and -2.

It equals one or the other, and you don't know which one.

SOLUTION NEXT PAGE

ALGEBRA

EXPRESSIONS

$x^2+8x+15$

DRILL — **FACTORING EXPRESSIONS**

Find the roots of these equations:

$x^2 + 8x = -15$ $\qquad\qquad$ $x^2 - 4x = -4$

$(x-3)(x-5)$ $\qquad$ $(x-2)(x-2)$

SOLUTION NEXT PAGE

DETAIL — **FACTORING EXPRESSIONS**

A perfect square minus a perfect square can be solved in two ways:

$$x^2 - 9 = 0$$

Factoring	**Square Root**
$(x-3)(x+3) = 0$	$x^2 = 9$
$x = 3$ or -3	$x = 3$ or -3

That's why a perfect square that equals a perfect square is called a **square root.**

DRILL — FACTORING EXPRESSIONS

Find the roots of these equations: $x^2 + 8x = -15$ and $x^2 - 4x = -4$

Steps	Examples	
1. Set them equal to zero	$x^2 + 8x + 15 = 0$	$x^2 - 4x + 4 = 0$
2. Put two sets of parentheses underneath the equation	$(\quad)(\quad) = 0$	$(\quad)(\quad) = 0$
3. Because $(x)(x) = x^2$, put x's in the parentheses, on the left.	$(x \quad)(x \quad) = 0$	$(x \quad)(x \quad) = 0$
4. Factor the numbers and put the factors in the parentheses, on the right.	$(x \quad 3)(x \quad 5) = 0$	$(x \quad 2)(x \quad 2) = 0$
5. Determine whether the numbers are positive or negative.	$(x + 3)(x + 5) = 0$	$(x - 2)(x - 2) = 0$
6. The roots of the equations are the values of x that make the equations equal zero.	$x = -3$ or -5	$x = 2$

QUIZ 1 FACTORING EXP

The second equation is an example of x^2 having only one possible value.

QUIZ — FACTORING EXPRESSIONS

If 3 is a root of the expression $ax^2 - 3x - 9 = 0$, what is the value of a?

(A) 2
(B) 3
(C) 5
(D) 7
(E) 9

SOLUTION NEXT PAGE

ALGEBRA | **QUIZ 1** | **FACTORING EXPRESSIONS**

FACTORING

If 3 is a root of the expression $ax^2 - 3x - 9 = 0$, what is the value of a?

(A) 2
(B) 3
(C) 5
(D) 7
(E) 9

SOLUTION

STEP 1 | **Separate the information from the question**

Copy down the expression $$ax^2 - 3x - 9 = 0$$

STEP 2 | **Turn this into something you can use**

Knowing that one of the roots is 3, substitute 3 for x in the equation. $$a3^2 - (3)(3) - 9 = 0$$

STEP 3 | **Use this to answer the question**

Solve for a

$$9a - 9 - 9 = 0$$

$$9a - 18 = 0$$
$$9a = 18$$
$$a = 2$$

Answer is A

If n, p, q, and r are positive integers such that $nq + nr + pq + pr = 35$, what is the value of $n + p + q + r$?

(A) 3
(B) 5
(C) 8
(D) 10
(E) 12

The **Quadratic Formula** is an alternate method to solve for x other than factoring the expressions.

$$x = \frac{-b \pm \sqrt{b^2 - 4ac}}{2a}$$

In all the years of teaching and reviewing the GMAT with students, I have never had to use this formula.

SOLUTION NEXT PAGE

ALGEBRA	QUIZ 2	FACTORING EXPRESSIONS

EXPRESSIONS

If n, p, q, and r are positive integers such that $nq + nr + pq + pr = 35$, what is the value of $n + p + q + r$?

(A) 3
(B) 5
(C) 8
(D) 10
(E) 12

SOLUTION

STEP 1 Separate the information from the question

Copy down the equation $nq + nr + pq + pr = 35$

STEP 2 Turn this into something you can use

When the arrangement doesn't make sense, look for a pattern.

Apparently, n and p are each multiplied by q and r.

$$(n + p)(q + r) = 35$$

Two expressions multiplied together equal 35.

One must equal 7, and the other must equal 5.

Having ... $(n + p)(q + r) = 35 \times 1$, making $n + p + q + r = 36$ isn't possible.

This is because n, p, q, and r are all positive integers, meaning each equals 1 or more.

Therefore, neither $(n + p)$ nor $(q + r)$ could equal 1.

$$(n + p)(q + r) = 7 \times 5$$

STEP 3 Use this to answer the question

$(n + p) = 7$ and $(q + r) = 5$, or vice versa,

but because they're added, it doesn't matter.

$$7 + 5 = 12$$

Answer is E

TOPIC FACTORING TERMS

ALGEBRA

EXPRESSIONS

Factoring terms is often the key to quickly answering even the most challenging GMAT question.

LIFELINE 67

When factoring terms, factor out the largest item that's in each term.

Steps	**Example**
1. To factor the expression	$9x + 6y - 3$
2. Find the largest item that's in each term.	3
3. Factor this from each term	$3(3x + 2y - 1)$
4. Check your work by multiplying it out	$3(3x + 2y - 1)$ $9x + 6y - 3$

DRILL FACTORING TERMS

Factor $2x^3 + 4x^2 + 6x$

$2(x^3+2x^2+3x)$

$2x^3+4x^2+6x$

QUIZ FACTORING TERMS

If $npq \neq 0$, then $\dfrac{n^2pq - np^2q - npq^2}{npq} =$

$\dfrac{npq(n-p-q)}{npq} = n-p-q$

(A) npq
(B) $n^2p^2q^2$
(C) $n^2 - p^2 - q^2$
(D) $(npq)^2$
(E) $n - p - q$

SOLUTION NEXT PAGE

ALGEBRA

EXPRESSIONS

DRILL — FACTORING TERMS

SOLUTION

Factor $2x^3 + 4x^2 + 6x$

$2x$ is the largest item in all three terms $2x(x^2 + 2x + 3)$

Check this by multiplying it out $2x(x^2 + 2x + 3)$

$2x^3 + 4x^2 + 6x$

QUIZ — FACTORING TERMS

$$\frac{n^2pq - np^2q - npq^2}{npq}$$

SOLUTION

STEP 1 **Separate the information from the question**

Copy down the expression $\frac{n^2pq - np^2q - npq^2}{npq}$

STEP 2 **Turn this into something you can use**

Find the largest item in each term. npq

Factor this from each term $\frac{npq(n - p - q)}{npq}$

STEP 3 **Use this to answer the question**

Cancel the npq from the fraction $\frac{\cancel{npq}(n - p - q)}{\cancel{npq}}$

The result is the answer $n - p - q$

Answer is E

Check your factoring by multiplying out the expression: $npq(n - p - q)$

$= n^2pq - np^2q - npq^2$

TOPIC | **FACTORING EXPONENTS** | **ALGEBRA**

EXPRESSIONS

Factoring exponents is the next part of the exponent lesson. For a refresher on Advanced Exponents, see Arithmetic, p. 91

LIFELINE 68

When factoring exponents, start with the smaller exponent.

Steps	**Example**
1. Copy down the expression	$x^6 - x^4$
2. Start with the smaller exponent	x^4
3. Factor this from each term	$x^4(x^2 - 1)$
Check your work	
4. Multiply out the expression	$x^4(x^2 - 1)$ $x^6 - x^4$

$x^6 - x^4$ is not x^2, though that's a common mistake.

DRILL | **FACTORING EXPONENTS**

1) Simplify $2^6 - 2^4$

2) Simplify $\dfrac{3^9 - 3^7}{3^8 - 3^6}$

SOLUTION NEXT PAGE

ALGEBRA	DRILL	FACTORING EXPONENTS

EXPRESSIONS

SOLUTION

1) Simplify $2^6 - 2^4$

1. Copy down the expression	$2^6 - 2^4$
2. Start with the smaller exponent	2^4
3. Factor this from each term	$2^4(2^2 - 1)$
4. Factor this from each term	$2^4(3)$

2) Simplify $\frac{3^9 - 3^7}{3^8 - 3^6}$

1. Copy down the expression	$\frac{3^9 - 3^7}{3^8 - 3^6}$
2. Use the smaller exponent from the top and bottom	$\frac{3^7(\text{top})}{3^6(\text{bottom})}$
3. Factor these from their respective terms	$\frac{3^7(3^2 - 1)}{3^6(3^2 - 1)}$
4. Cancel like terms	$\frac{3^7\cancel{(3^2 - 1)}}{3^6\cancel{(3^2 - 1)}}$
5. Reduce the exponents	$\frac{3^7}{3^6} = 3$

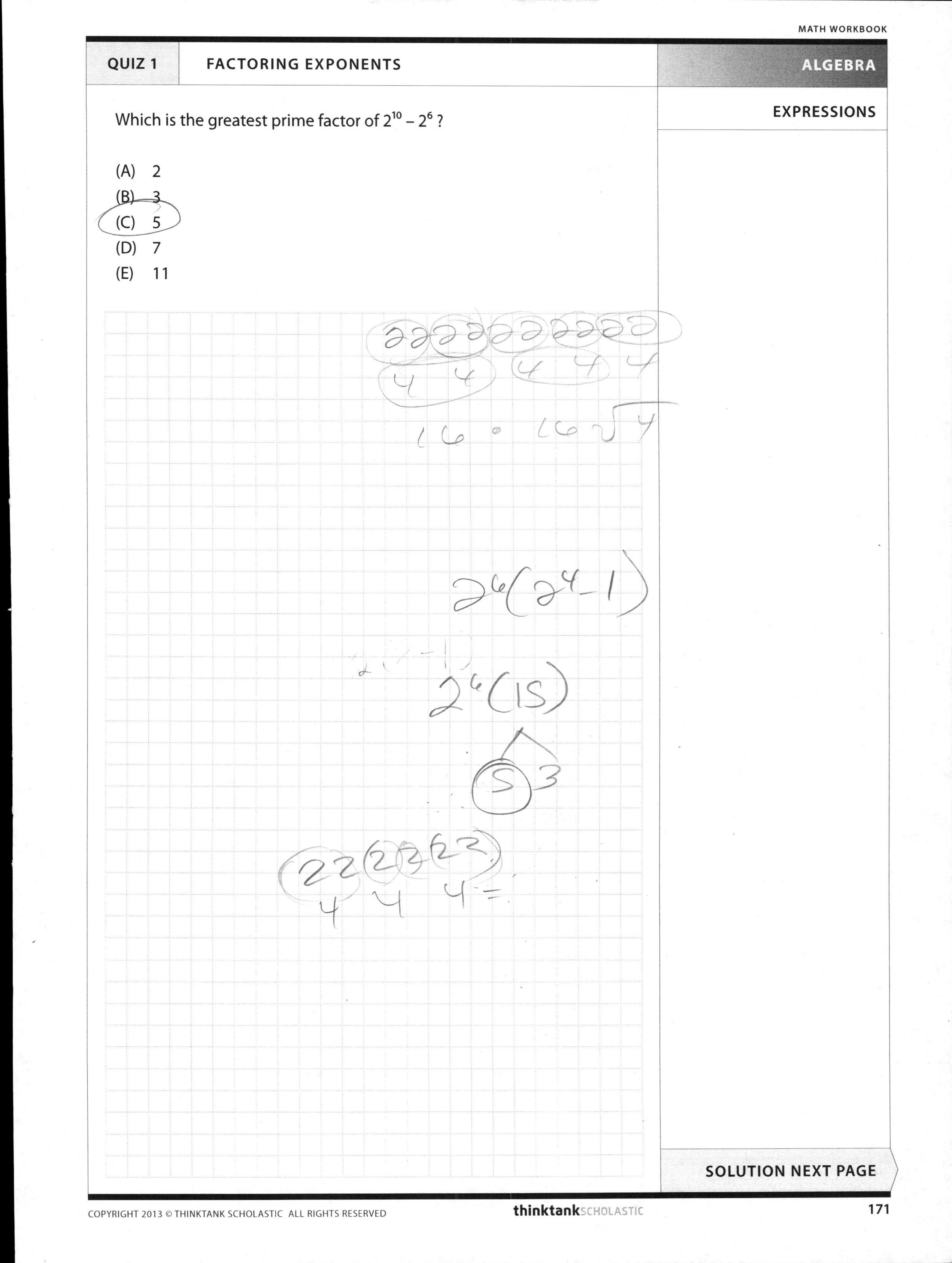

Which is the greatest prime factor of $2^{10} - 2^{6}$?

(A) 2
(B) 3
(C) 5
(D) 7
(E) 11

SOLUTION NEXT PAGE

ALGEBRA | **QUIZ 1** | **FACTORING EXPONENTS**

EXPRESSIONS

Which is the greatest prime factor of $2^{10} - 2^{6}$?

(A) 2
(B) 3
(C) 5
(D) 7
(E) 11

SOLUTION

STEP 1 Separate the information from the question

Copy down the expression $2^{10} - 2^{6}$

STEP 2 Turn this into something you can use

Use the smaller exponent 2^{6}

Factor this from each term $2^{6}(2^{4} - 1)$

Simplify $(2^{4} - 1)$ into $(16 - 1)$ then (15) $2^{6}(15)$

STEP 3 Use this to answer the question

To find the prime factors, $2^{6}(15)$

Factor the expression to its primes $2 \times 2 \times 2 \times 2 \times 2 \times 2 \times 3 \times 5$

Find the largest prime factor 5

Answer is C

If $3^h + 3^{h-2} = 10(3^9)$, then $h =$

(A) 7
(B) 9
(C) 11
(D) 13
(E) 15

SOLUTION NEXT PAGE

ALGEBRA	QUIZ 2	FACTORING EXPONENTS

EXPRESSIONS

If $3^h + 3^{h-2} = 10(3^9)$, then $h =$

(A) 7
(B) 9
(C) 11
(D) 13
(E) 15

SOLUTION

STEP 1 Separate the information from the question

Copy down the equation $3^h + 3^{h-2} = 10(3^9)$

STEP 2 Turn this into something you can use

1. Factor the left side $3^h + 3^{h-2}$
2. Use the smaller exponent 3^{h-2}
3. Factor this from each term $3^{h-2}(3^2 + 1)$
4. Simplify the expression $(3^2 + 1)$ $3^{h-2}(10)$

$h - 2$ is smaller than h

STEP 3 Use this to answer the question

1. Recopy the equation $3^{h-2}(10) = 10(3^9)$
2. Cancel the 10's $3^{h-2} = 3^9$
3. Eliminate the 3's $h - 2 = 9$
4. Solve for h $h = 11$

For 3^{h-2} to equal 3^9, the exponents have to be the same, meaning $h - 2 = 9$.

Answer is C

TOPIC **MATH OPERATORS**

Basic math operators are these:

$+ \quad - \quad \div \quad \times$

The GMAT invents its own operators, with unique symbols:

$\blacksquare \quad \odot \quad \blacktriangle$

Fortunately, the GMAT also tells you what the symbol means.

For example: $x \odot y = xy + 3(x - y)$

And the equation is simple.

LIFELINE 69

When working with new math operators, use real numbers for *x* and *y*.

For example: $2 \odot 3 =$

Steps	**Example**
1. Copy the equation down with the operator	$x \odot y = xy + 3(x - y)$
2. Substitute (in this example) 2 for *x* and 3 for *y* in the equation	$2 \odot 3 = (2)(3) + 3(2 - 3)$
3. Simplify the equation	$2 \odot 3 = 6 + 3(-1)$
	$= 6 + -3$
	$= 3$

DRILL **MATH OPERATORS**

Using the same equation

$x \odot y = xy + 3(x - y)$

Solve:

$5 \odot 4 =$

(5)(4)+3(5−4)
20+15−12
35
−12
23

SOLUTION NEXT PAGE

ALGEBRA | **DRILL** | **MATH OPERATORS**

MATH OPERATORS

Using the same equation

$x \odot y = xy + 3(x - y)$

$5 \odot 4 =$

SOLUTION

The math will always be simple. Understanding the question is the challenge.

Steps	Example
1. Copy down the equation with the operator	$x \odot y = xy + 3(x - y)$
2. Substitute 5 for *x* and 4 for *y* in the equation	$5 \odot 4 = (5)(4) + 3(5 - 4)$
3. Simplify the equation	$5 \odot 4 = 20 + 3(1)$
	$5 \odot 4 = 23$

QUIZ 1 **MATH OPERATORS** **ALGEBRA**

MATH OPERATORS

For all numbers x and y, the operation ✪ is defined by x ✪ $y = x^2 + y^2 - 1$

What is the value of 4 ✪ 5 ?

(A) 8
(B) 9
(C) 10
(D) 40
(E) 80

4 ✪ 5
4 = x 5 = y

SOLUTION NEXT PAGE

ALGEBRA | QUIZ 1 | MATH OPERATORS

MATH OPERATORS

For all numbers x and y, the operation ✪ is defined by x ✪ $y = x^2 + y^2 - 1$

What is the value of 4 ✪ 5 ?

(A) 8
(B) 9
(C) 10
(D) 40
(E) 80

SOLUTION

STEP 1 **Separate the information from the question**

Copy down the equation with the operator x ✪ $y = x^2 + y^2 - 1$

STEP 2 **Turn this into something you can use**

Substitute 4 for x and 5 for y in the equation 4 ✪ 5 $= 4^2 + 5^2 - 1$

STEP 3 **Use this to answer the question**

Simplify the equation 4 ✪ 5 $= 16 + 25 - 1$

4 ✪ 5 $= 40$

Answer is D

QUIZ 2 MATH OPERATORS

MATH OPERATORS

For all positive integers a, $°a = 2a$ when a is even and $°c = 3a$ when a is odd. Which of the following is equivalent to $°5 + °6$?

(A) °9
(B) °11
(C) °18
(D) °27
(E) °33

SOLUTION NEXT PAGE

ALGEBRA	QUIZ 2	MATH OPERATORS

MATH OPERATORS

For all positive integers a, $^\circ a = 2a$ when a is even and $^\circ a = 3a$ when a is odd. Which of the following is equivalent to $^\circ 5 + {}^\circ 6$?

(A) $^\circ 9$
(B) $^\circ 11$
(C) $^\circ 18$
(D) $^\circ 27$
(E) $^\circ 33$

SOLUTION

STEP 1 Separate the information from the question

$^\circ a = 2a$ when a is even

$^\circ a = 3a$ when a is odd

STEP 2 Turn this into something that makes sense

when a is even, such as 6,	when a is odd, such as 5,
$^\circ a = 2a$	$^\circ a = 3a$
$^\circ 6 = 2 \times 6$	$^\circ 5 = 3 \times 5$
$^\circ 6 = 12$	$^\circ 5 = 15$

STEP 3 Use this to answer the question

$^\circ 5 + {}^\circ 6 = 15 + 12 \quad = 27$

D is the trap answer.

Though you're looking for 27, D is actually $^\circ 27$, which equals 3×27.

A, on the other hand, is $^\circ 9$. $\quad ^\circ 9 = 3 \times 9 = 27$

Answer is A

Devious? Yes. Like the GMAT? Absolutely.

Be sure to stay on your toes and watch for traps, no matter how comfortable you become with the math.

SEQUENCES

A **sequence** is a repeating letter with small numbers:

$$a_1, a_2, a_3, \ldots a_n$$

It could be any letter, not just *a*.

The question looks like this.

> If $a_1, a_2, a_3, \ldots a_n$ is a sequence of positive numbers where $a_1 = 5$ and $a_{n+1} = a_n + 3$, what is the value of a_6?

The small number is a counter.

Think of the small number as a locker number, and the large number as its contents.

The locker number has nothing to do with the contents.

For example: a_3 represents locker #3.
Locker #3 doesn't have a 3 inside of it.

ALGEBRA

SEQUENCES

LIFELINE 70

When working a sequence, find the pattern.

Example: If $a_1, a_2, a_3, \ldots a_n$ is a sequence of positive numbers where $a_1 = 5$ and $a_{n+1} = a_n + 3$, what is the value of a_6?

Steps	Example
1. List about five a_n's	a_1 a_2 a_3 a_4 a_5
2. Put down any values for a_n that you have	$a_1 = 5$ a_2 a_3 a_4 a_5
3. Interpret the equation from into or in English,	$a_{n+1} = a_n + 3$ $a_2 = a_1 + 3$ "the next one is 3 more"

Though the value of n changes, it always has one value at one time.
If $n = 1$, it's the same on both sides of the equation.

4. Fill in the rest of the a_n's for the answer.
If locker #1 has a 5, then according to the equation, locker #2 has that plus 3, or 8. Write that down.

$a_1 = 5$
$a_2 = 8$

5. Locker #3 has 3 more than locker #2, for 11. Write that down.

$a_3 = 11$

The pattern is found, so you no longer need the equation.

$a_4 = 14$
$a_5 = 17$
$a_6 = 20$

In the arithmetic sequence $g_1, g_2, g_3, \ldots g_n \ldots, g_2 = 8$ and $g_{n+1} = g_n + 7$. What is the value of n when $g_n = 50$?

(A) 5
(B) 6
(C) 7
(D) 8
(E) 9

SOLUTION NEXT PAGE

ALGEBRA

SEQUENCES

QUIZ | SEQUENCES

In the arithmetic sequence $g_1, g_2, g_3, \ldots g_n \ldots, g_2 = 8$ and $g_{n+1} = g_n + 7$. What is the value of n when $g_n = 50$?

(A) 5
(B) 6
(C) 7
(D) 8
(E) 9

SOLUTION

STEP 1 | **Separate the information from the question**

Copy down the equation $\quad g_{n+1} = g_n + 7$

STEP 2 | **Turn this into something you can use**

1. List about five g_n's

g_1
g_2
g_3
g_4
g_5

2. Put down any values for g_n that you have

g_1
$g_2 = 8$
g_3
g_4
g_5

3. Interpret the equation

from $g_{n+1} = g_n + 7$

into $g_3 = g_2 + 7$

or in English, "the next one is 7 more"

4. Fill in the rest of the g_n's for the answer.

g_1
$g_2 = 8$
$g_3 = 15$
$g_4 = 22$
$g_5 = 29$

STEP 3 | **Use this to answer the question**

To reach 50, continue the list.

$g_6 = 36$
$g_7 = 43$
$g_8 = 50$

n represents the counter. When $g_n = 50$, $n = 8$.

Answer is D

WORKING WITH UNKNOWNS

Working with unknowns means you don't find a numeric answer. You find an answer in terms of a letter, such as *x*.

For example, if every day I haul *r* cans to the recycle bin, then in 5 days, I hauled 5*r* cans. You don't actually find out the number of cans, but the answer is in terms of *r*.

LIFELINE 71

When working with unknowns, use real numbers to set up the equation, but stick with letters to do the work.

Example: If *P* people agreed to split the cost of *L* lunch, then in terms of *P* and *L*, what was the cost per person?

Steps	**Example**
1. Use numbers to set up the equation.	If 10 people split the cost of a $100 lunch
The cost per person is	$\frac{\$100}{10}$
2. Stick with letters to do the work.	$\frac{L}{P}$

Don't simplify the $\frac{\$100}{10}$ to $10 per person because that won't help you place the $\frac{L}{P}$

DRILL — WORKING WITH UNKNOWNS

Bobby hit *h* more home runs than Joey, and together they hit a total of *r* home runs. In terms of *h* and *r*, what represents the number of home runs that Bobby hit?

SOLUTION NEXT PAGE

WORKING WITH UNKNOWNS

Bobby hit h more home runs than Joey, and together they hit a total of r home runs. In terms of h and r, what represents the number of home runs that Bobby hit?

SOLUTION

STEP 1 Separate the information from the question

Bobby hit h more home runs than Joey.

Together they hit a total of r home runs

STEP 2 Turn this into something you can use

Set up the equation with numbers.

Bobby hit 5, Joey hit 3, together they hit 8. $5 + 3 = 8$

Find the r: $B + J = r$

or B (5) hit h (2) more than J (3) $2 = 5 - 3$

Find the h: $h = B - J$

STEP 3 Use this to answer the question

Stick with letters to do the work

$$h = B - J$$

$$B + J = r$$

To solve for B, or Bobby, combine the equations to remove J.

This is the **subtraction** method for simultaneuos equations. Here the equations are added, because J is already negative.

$$h = B - J$$

$$+r = B + J$$

$$h + r = 2B + 0$$

$$h + r = 2B$$

$$\frac{h + r}{2} = B$$

In terms of h and r, Bobby hit $\frac{h + r}{2}$ home runs.

To celebrate graduation, the S students in a class agreed to share equally the cost of renting a yacht. If the yacht costs a total of Y dollars, and T additional students joined and agreed to equally share the cost, which of the following represents the reduced amount, in dollars, that each of the original students would have to pay so that the cost of the yacht is completely paid, and not overpaid?

(A) $\frac{Y}{S}$ (B) $\frac{Y}{S+T}$ (C) $\frac{Y}{S-T}$ (D) $\frac{YT}{S-T}$ (E) $\frac{YT}{S(S+T)}$

WORKING WITH UNKNOWNS

SOLUTION NEXT PAGE

ALGEBRA | QUIZ | WORKING WITH UNKNOWNS

WORKING WITH UNKNOWNS

To celebrate graduation, the S students in a class agreed to share equally the cost of renting a yacht. If the yacht costs a total of Y dollars, and T additional students joined and agreed to equally share the cost, which of the following represents the reduced amount, in dollars, that each of the original students would have to pay so that the cost of the yacht is completely paid, and not overpaid?

SOLUTION

STEP 1 Separate the information from the question

S = number of original students
Y = cost of the yacht
T = number of additional students

STEP 2 Turn this into something you can use

- Write an equation of the relationship — $S + T$ = the new student total
- Use numbers to figure out the relationship — If the yacht costs \$60, and there were 4 original students, then the original cost for each student is $\frac{\$60}{4}$
- Stick with letters to do the work — $\frac{\$60}{4} = \frac{Y}{S}$ = Original student cost

If 2 students joined, then the new cost per student is $\frac{\$60}{6}$

$\frac{\$60}{6} = \frac{Y}{S+T}$ = New student cost

STEP 3 Use this to answer the question

If the original cost is $\frac{\$60}{4}$ and the new cost is $\frac{\$60}{6}$, then the reduced amount per student is $\frac{\$60}{4} - \frac{\$60}{6}$ or $\frac{Y}{S} - \frac{Y}{S+T}$

1. Find the common denominator
2. Subtract the numerator

$$\frac{Y(S+T)}{S(S+T)} - \frac{S(Y)}{S(S+T)} = \frac{Y(S+T) - S(Y)}{S(S+T)}$$

3. Simplify the numerator
   a. Distribute the Y's
   b. Cancel the SY's

$$\frac{SY + TY - SY}{S(S+T)} = \frac{TY}{S(S+T)}$$

Answer is E

TOPIC: SIMPLIFYING FUNCTIONS

ALGEBRA

***f*(*x*) FUNCTIONS**

$f(x)$ is a term borrowed from calculus and refers to the **function** of x.

LIFELINE 72

When solving $f(x)$ functions, treat the $f(x)$ like a y.

Start with an equation: $y = x^2 + 2$

Pick a value for x, — If $x = 2$,

and get a value for y. — $y = 6$

The GMAT uses $f(x)$ instead of y: $y = x^2 + 2 \longrightarrow f(x) = x^2 + 2$

It works exactly the same way.

Using the same equation: $f(x) = x^2 + 2$

Pick a value for x, — If $x = 3$,

and get a value $f(x)$. — $f(3) = 3^2 + 2$

The value of $f(x)$ is written as: — $f(3) = 11$

DRILL: SIMPLIFYING FUNCTIONS

If $f(x) = x^2 + 2$, solve for:

1) $f(5)$

2) $f(z)$

3) $\dfrac{1}{f(3)}$

4) $\dfrac{1}{f(x)}$

5) $\dfrac{1}{f(h)} \times \dfrac{1}{f(2)}$

6) $f(2x)$

7) $2f(x)$

SOLUTION NEXT PAGE

ALGEBRA | **DRILL** | **SIMPLIFYING FUNCTIONS**

$f(x)$ FUNCTIONS

If $f(x) = x^2 + 2$

SOLUTION

1) $f(5) = 5^2 + 2$
$= 27$

Always replace the x in the equation with what's in the parentheses.

2) $f(z) = z^2 + 2$

Solve for $f(3)$ first, then place it back into the fraction.

3) $f(3) = 3^2 + 2$
$= 11$
$\frac{1}{f(3)} = \frac{1}{11}$

4) $f(x) = x^2 + 2$
$\frac{1}{f(x)} = \frac{1}{x^2 + 2}$

5) $\frac{1}{f(h)} \text{ x } \frac{1}{f(2)}$

$\frac{1}{f(h)} = \frac{1}{h^2 + 2}$ and $\frac{1}{f(2)} = \frac{1}{6}$

$\frac{1}{h^2 + 2} \times \frac{1}{6} = \frac{1}{6h^2 + 12}$

6) $f(2x) = (2x)^2 + 2$
$= 4x^2 + 2$

7) $f(x) = x^2 + 2$
$2f(x) = 2x^2 + 4$

For all numbers y such that $y \neq 1$, if $f(y) = \dfrac{y^2+2}{y+1}$, then $\left(\dfrac{1}{f(y)}\right)\left(\dfrac{1}{f(2)}\right) =$

(A) $\dfrac{y+1}{2(y^2+2)}$

(B) $\dfrac{2(y+1)}{y^2+2}$

(C) $\dfrac{y^2+2}{2(y+1)}$

(D) $\dfrac{2(y^2+2)}{y+1}$

(E) $\dfrac{2(y^2+1)}{y+2}$

SOLUTION NEXT PAGE

ALGEBRA | **QUIZ** | **SIMPLIFYING FUNCTIONS**

$f(x)$ FUNCTIONS

For all numbers y such that $y \neq 1$, if $f(y) = \frac{y^2+2}{y+1}$, then $\left(\frac{1}{f(y)}\right)\left(\frac{1}{f(2)}\right) =$

(A) $\frac{y+1}{2(y^2+2)}$

(B) $\frac{2(y+1)}{y^2+2}$

(C) $\frac{y^2+2}{2(y+1)}$

(D) $\frac{2(y^2+2)}{y+1}$

(E) $\frac{2(y^2+1)}{y+2}$

SOLUTION

STEP 1 **Separate the information from the question**

$$f(y) = \frac{y^2+2}{y+1}$$

STEP 2 **Turn this into something you can use**

It's not always the letters f and x; for example, it could be $g(h)$. Regardless of the letters, the equation works exactly the same way.

To find $f(2)$, substitute 2 for y in the fraction

$$\frac{y^2+2}{y+1} \rightarrow \frac{2^2+2}{2+1} = \frac{6}{3} = 2$$

$f(2) = 2$

STEP 3 **Use this to answer the question**

Remember
A "1" over any fraction flips that fraction.

If $f(2) = 2$, then $\frac{1}{f(2)} = \frac{1}{2}$

If $f(y) = \frac{y^2+2}{y+1}$, then $\frac{1}{f(y)}$ is the reciprocal: $\frac{y+1}{y^2+2}$

Multiply these fractions

$$\left(\frac{y+1}{y^2+2}\right)\left(\frac{1}{2}\right) = \frac{y+1}{2(y^2+2)}$$

Answer is A

TOPIC FINDING THE $f(x)$ FUNCTION

ALGEBRA

$f(x)$ FUNCTIONS

The GMAT varies the $f(x)$ question by giving you the answer and asking you to find the function. The answer choices are the functions, and you have to try each one.

LIFELINE 73

When finding the $f(x)$ function, use real numbers for x.

Example: If $x = 7$, for which function does $f(x) = 12$?

(A) $f(x) = x + 7$
(B) $f(x) = x^2$
(C) $f(x) = 2x - 2$

Steps	**Example**
1. Because $x = 7$, replace x with 7 in each answer choice	(A) $f(7) = 7 + 7$ (B) $f(7) = 7^2$ (C) $f(7) = 2(7) - 2$
2. Solve for $f(7)$ in each answer choice	(A) $f(7) = 14$ (B) $f(7) = 49$ (C) $f(7) = 12$

Answer is C, where $f(7) = 12$

DRILL FINDING THE $f(x)$ FUNCTION

1. If $x = 5$, for which equation does $y = 3$?

(A) $y = x + 2$
(B) $y = x^2$
(C) $y = x - 2$

2. For which function does $f(x) = 2x$?

(A) $f(x) = x + 2$
(B) $f(x) = 3x - x$
(C) $f(x) = x^2$

SOLUTION NEXT PAGE

ALGEBRA

$f(x)$ FUNCTIONS

DRILL

FINDING THE $f(x)$ FUNCTION

1. If $x = 5$, for which equation does $y = 3$?

(A) $y = x + 2$
(B) $y = x^2$
(C) $y = x - 2$

SOLUTION

These are simple examples, but they show the nature of the question.

Working the math is always easy; understanding the question is the challenge.

Steps	Example
Because $x = 5$, replace x with 5 in each answer choice	(A) $y = 5 + 2$ (B) $y = 5^2$ (C) $y = 5 - 2$
Solve for y in each answer choice	(A) $y = 7$ (B) $y = 25$ (C) $y = 3$

Answer is C

2. For which function does $f(x) = 2x$?

(A) $f(x) = x + 2$
(B) $f(x) = 3x - x$
(C) $f(x) = x^2$

Steps	Example
Simplify the value of $f(x)$ in each answer choice	(A) $f(x) = x + 2$ (B) $f(x) = 2x$ (C) $f(x) = x^2$

Answer is B

QUIZ | FINDING THE $f(x)$ FUNCTION | ALGEBRA

$f(x)$ FUNCTIONS

For which of the following functions is $f(x) = f\left(\frac{1}{x}\right)$ for all numbers x?

(A) $f(x) = \frac{1}{x}$

(B) $f(x) = \frac{1}{x^2}$

(C) $f(x) = x - 1$

(D) $f(x) = x^2$

(E) $f(x) = x + \frac{1}{x}$

SOLUTION NEXT PAGE

ALGEBRA | **QUIZ** | **FINDING THE $f(x)$ FUNCTION**

$f(x)$ FUNCTIONS

For which of the following functions is $f(x) = f\left(\frac{1}{x}\right)$ for all numbers x?

(A) $f(x) = \frac{1}{x}$

(B) $f(x) = \frac{1}{x^2}$

(C) $f(x) = x - 1$

(D) $f(x) = x^2$

(E) $f(x) = x + \frac{1}{x}$

Though the value of x changes, it always has one value at one time. Both x's are the same.

SOLUTION

STEP 1 Separate the information from the question

Where does $f(x)$ equal $f\left(\frac{1}{x}\right)$?

STEP 2 Turn this into something you can use

Pick a number for x, such as 2.

Where does $f(2)$ equal $f\left(\frac{1}{2}\right)$?

STEP 3 Use this to answer the question

Check each answer choice.

	$f(2)$	$f\left(\frac{1}{2}\right)$	Do they equal each other?
(A) $f(x) = \frac{1}{x}$	$\frac{1}{2}$	2	No, $\frac{1}{2}$ doesn't equal 2
(B) $f(x) = \frac{1}{x^2}$	$\frac{1}{4}$	4	No
(C) $f(x) = x - 1$	1	$-\frac{1}{2}$	No
(D) $f(x) = x^2$	4	$\frac{1}{4}$	No
(E) $f(x) = x + \frac{1}{x}$	$2\frac{1}{2}$	$2\frac{1}{2}$	Yes

Whenever trying all the answer choices, work the first one, then the last one, then work your way up.

Answer is E

TOPIC	TRYING ANSWER CHOICES	ALGEBRA

ODDBALL QUESTIONS

The GMAT features **oddball questions** in its mix. Each has a unique solving method that doesn't apply to any other question, and it's impossible to anticipate all of them.

Like any other question, this is a puzzle, not a math problem, and the approach is key to answering it. You may go through a few trial-and-error solutions until you find one that works.

LIFELINE 74

When trying all the answer choices, work the first one, then the last one, then work your way up.

Sometimes all you can do is try the answer choices. Fortunately, there are only five, and the math is always simple.

QUIZ	TRYING ANSWER CHOICES

It's ok if you miss an early question—but not if you take ten minutes to struggle with it first.

If you have no idea what to do, guess and move on: you can still get a high score.

If $\begin{array}{r} AA \\ +\,BB \\ \hline AAC \end{array}$, what is $\frac{A}{B}$?

(A) $\frac{5}{4}$

(B) $\frac{4}{6}$

(C) $\frac{3}{7}$

(D) $\frac{2}{8}$

(E) $\frac{1}{9}$

SOLUTION NEXT PAGE

ALGEBRA | QUIZ | TRYING ANSWER CHOICES

ODDBALL QUESTIONS

If $\begin{array}{r} AA \\ +BB \\ \hline AAC \end{array}$, what is $\frac{A}{B}$?

SOLUTION

STEP 1 Separate the information from the question

$$\begin{array}{r} AA \\ +BB \\ \hline AAC \end{array}$$

STEP 2 Turn this into something you can use

You know that *A*, *B*, and *C* represent numbers, because you have to choose a numeric fraction to represent $\frac{A}{B}$

So without checking the math, the fraction is something like $\begin{array}{r} 55 \\ +66 \\ \hline 557 \end{array}$

STEP 3 Use this to answer the question

Knowing the letters represent numbers, and there are only five answers, try each one.

Plug the answer choices in for the *A*'s and *B*'s that are added, and look for a result where the *A*'s match the sum.

To match this question	**Try this** first answer choice of $\frac{5}{4}$ for $\frac{A}{B}$
$\begin{array}{r} AA \\ +BB \\ \hline AAC \end{array}$	$\begin{array}{r} 55 \\ +44 \\ \hline 99 \end{array}$

The top two numbers (*AA*) have to match the first two numbers in the bottom (*AAC*). In this case, they don't, so choose another.

Now try the last answer choice.

To match this equation	**Try this** last answer choice of $\frac{1}{9}$ for $\frac{A}{B}$
$\begin{array}{r} AA \\ +BB \\ \hline AAC \end{array}$	$\begin{array}{r} 11 \\ +99 \\ \hline 110 \end{array}$

Here the *A*'s match up (they're all 1's)

Answer is E

thinktankSCHOLASTIC

TOPIC **SIMPLIFYING THE QUESTION** | ALGEBRA

ODDBALL QUESTIONS

The GMAT challenges you to be creative and resourceful. With a complex problem, instead of jumping into the math, look for the pattern by simplifying the math.

LIFELINE 75

When up against complex math, use simple math to find the pattern.

Note: 5! refers to the **factorial** of 5, which is the product of 5 and all the integers less than it.

$$5! = 5 \times 4 \times 3 \times 2 \times 1 = 120$$

For more on the factorial, see Word Problems, p. 303.

QUIZ **SIMPLIFYING THE QUESTION**

If p is the smallest prime factor of $25! + 1$, then p is

(A) less than 5
(B) between 5 and 10
(C) between 10 and 15
(D) between 15 and 20
(E) greater than 20

SOLUTION NEXT PAGE

ALGEBRA | **QUIZ 1** | **SIMPLIFYING THE QUESTION**

ODDBALL QUESTIONS

If p is the smallest prime factor of 25! + 1, then p is

(A) less than 5
(B) between 5 and 10
(C) between 10 and 15
(D) between 15 and 20
(E) greater than 20

SOLUTION

STEP 1 Separate the information from the question

Find the smallest prime factor of 25! + 1

STEP 2 Turn this into something you can use

$(25 \times 24 \times 23 \ldots 3 \times 2 \times 1) + 1$

STEP 3 Use this to answer the question

Use simple math to find the pattern.

Find the smallest prime factor of 4! + 1
$(4 \times 3 \times 2 \times 1) + 1 = 25$
The smallest prime factor of 25 is 5.

Try another one:

Find the smallest prime factor of 5! + 1
$(5 \times 4 \times 3 \times 2 \times 1) + 1 = 121$
The smallest prime factor of 121 is 11.

Apparently, when you take the factorial of a number and add 1, you get a new number where the smallest prime factor is larger than the original number.

Try a third one, just to be sure:

Finding smallest prime factor of 6! + 1 is too much math.

Instead, find the smallest prime factor of 3! + 1
$(3 \times 2 \times 1) + 1 = 7$

The smallest prime factor of 7 is 7, which is greater than 3.

> **Remember**
> Every number is a factor of itself.

Based on this pattern, the smallest prime factor of 25! + 1 is probably greater than 25.

Answer is E

HOMEWORK | ALGEBRA

13th Editon *GMAT Review Book,* pp. 152-185, question numbers | GMAT REVIEW

ANSWERS NEXT PAGE

Absolute Value
14 ____
24 ____
51 ____
143 ____

Inequaltities
50 ____
73 ____
130 ____
154 ____
192 ____
229 ____

Coordinate Geometry
202 ____
206 ____

Linear Equations
28 ____
43 ____
61 ____
211 ____

Parabolas
7 ____

Simultaneous Equations
47 ____
55 ____
64 ____
83 ____
93 ____
102 ____
119 ____
123 ____
137 ____
140 ____
153 ____
167 ____

Multiplying Expressions
37 ____

Factoring Expressions
45 ____
99 ____
191 ____
223 ____

Math Operators
110 ____
134 ____
149 ____
156 ____

Functions
129 ____

Sequences
100
126
160 ____
190 ____

Working With Unknowns
4 ____
34 ____
54 ____
89 ____
91 ____
131 ____
177 ____
181 ____
184 ____
198 ____
205 ____
214 ____
218

ALGEBRA

GMAT REVIEW

ANSWERS

HOMEWORK

13th Editon *GMAT Review Book,* pp. 152-185, question numbers

Absolute Value
14 ____C
24 ____A
51 ____B
143 ____E

Inequaltities
50 ____B
73 ____A
130 ____D
154 ____C
192 ____E
229 ____D

Coordinate Geometry
202 ____D
206 ____B

Linear Equations
28 ____E
43 ____B
61 ____B
211 ____B

Parabolas
7 ____E

Simultaneous Equations
47 ____D
55 ____D
64 ____B
83 ____D
93 ____E
102 ____D
119 ____B
123 ____C
137 ____B
140 ____A
153 ____D
167 ____D

Multiplying Expressions
37 ____D

Factoring Expressions
45 ____A
99 ____B
191 ____C
223 ____C

Math Operators
110 ____B
134 ____E
149 ____A
156 ____A

Functions
129 ____B

Sequences
100 ____C
126 ____D
160 ____C
190 ____E

Working With Unknowns
4 ____C
34 ____B
54 ____B
72 ____C
89 ____A
91 ____A
131 ____C
177 ____D
181 ____E
184 ____A
198 ____D
205 ____E
214 ____D
218 ____B

Geometry

Geometry

Lesson 3

Geometry is the study of lines, angles, and shapes.

It all begins with the **line**.

Two lines make an **angle**.

Three or more lines make a shape.

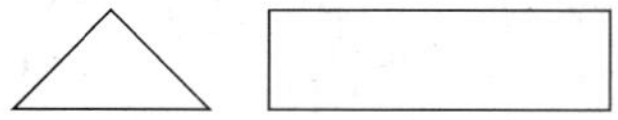

A line is straight and continuous.

If it curves, it's not a line.

A **segment** is part of a line,
usually offset by two points.

TOPIC	KNOWING ALL THE ANGLES	GEOMETRY

LINES AND ANGLES

Two crossing lines create four angles.

LIFELINE 76

With the measure of one angle, you can find the measures of the other three angles.

1) **Opposite** angles are equal

2) **Supplementary** angles total 180°

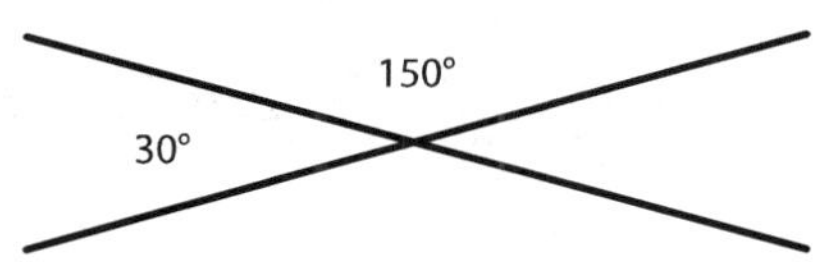

Therefore, if you know one angle, you automatically know all the angles.

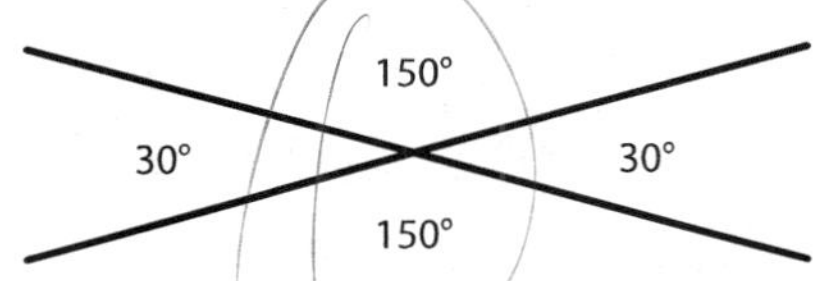

QUIZ 1 LINES AND ANGLES

In the figure shown, what is the value of x ?

(A) 140
(B) 150
(C) 180
(D) 200
(E) 220

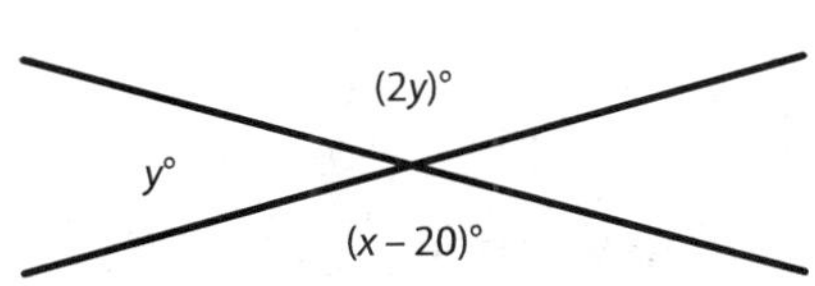

SOLUTION NEXT PAGE

GEOMETRY	QUIZ	LINES AND ANGLES

LINES AND ANGLES

In the figure shown, what is the value of x ?

(A) 140
(B) 150
(C) 180
(D) 200
(E) 220

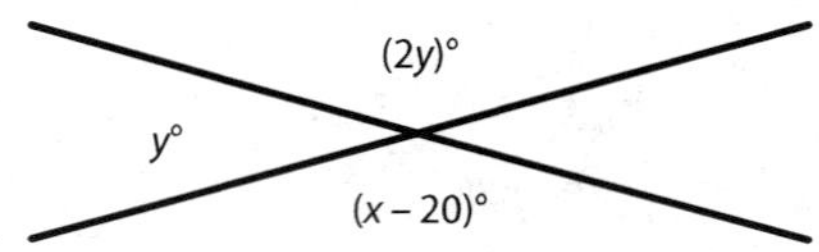

SOLUTION

STEP 1 Separate the information from the question

Copy down the drawing on your scratch paper

STEP 2 Turn this into something you can use

Supplementary angles total 180°, making this true: $y + 2y = 180$

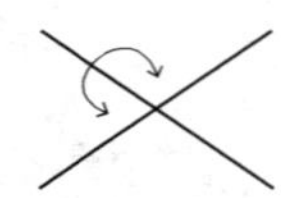

Opposite angles are equal, making this true: $x - 20 = 2y$

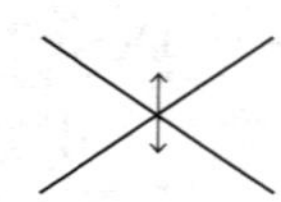

STEP 3 Use this to answer the question

1. Solve for y

$$3y = 180$$
$$y = 60$$

2. Substitute 60 for y in the second equation

$$x - 20 = 2y$$
$$x - 20 = 2(60)$$
$$x - 20 = 120$$

3. Solve for x

$$x = 140$$

Answer is A

A **polygon** is a shape with more than two sides, and all the sides are straight.

The terms **regular** and **equilateral** mean the same thing: all the sides and angles are the same.

Regular and Equilateral

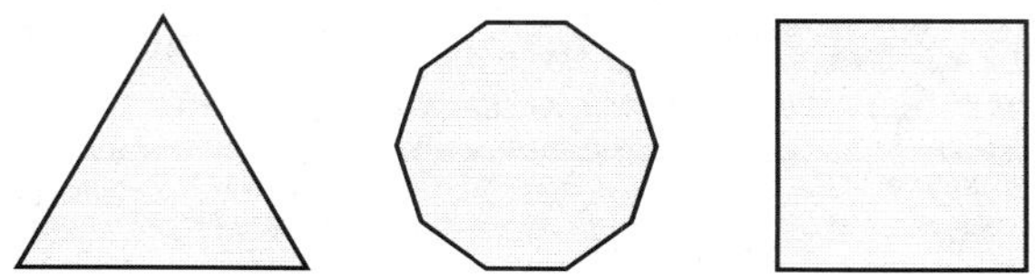

3 equal sides
3 equal angles

10 equal sides
10 equal angles

4 equal sides
4 equal angles

The term **equiangular** means the angles, but not the sides, are the same, as in a rectangle.

Equiangular

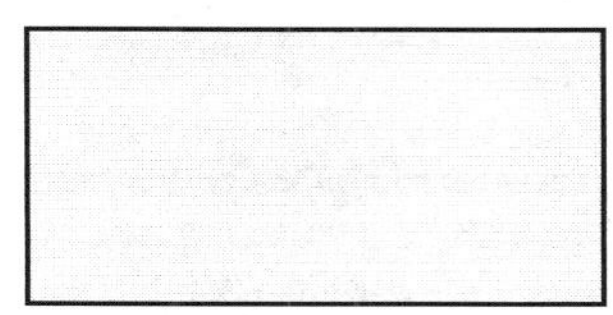

angles equal
sides not equal

The term **congruent** means the shapes are identical, even if one is rotated.

Congruent

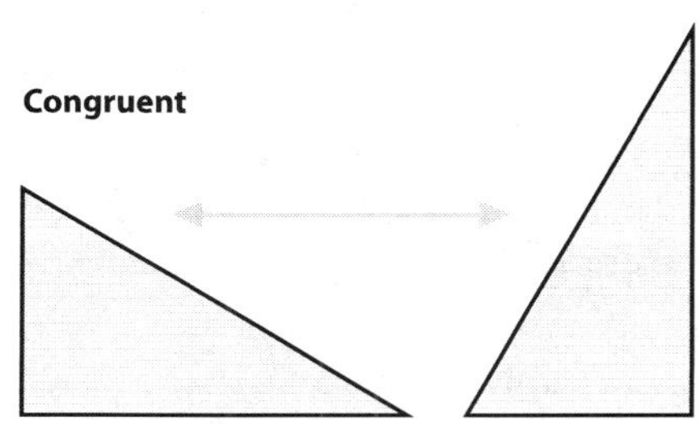

shapes are identical

The term **similar** means the angles and linear ratios are the same, but the sizes are different.

Similar

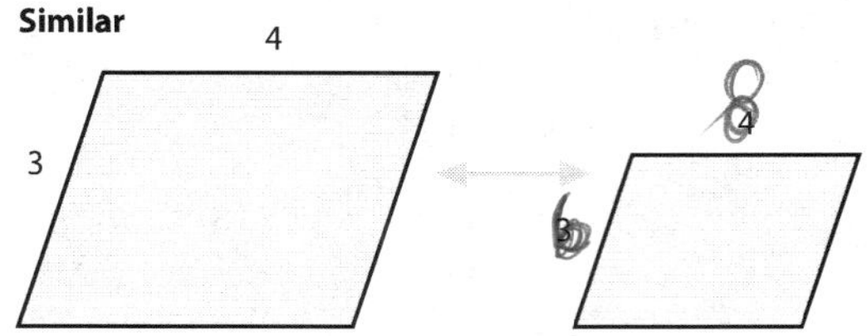

angles same
linear ratios same

size different

GEOMETRY

POLYGONS

LIFELINE 77

To find the angle totals, start by counting the angles.

The number of angles is the same as the number of sides.

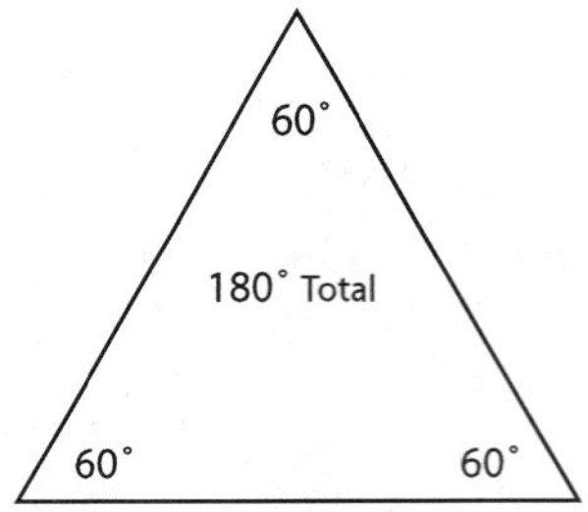

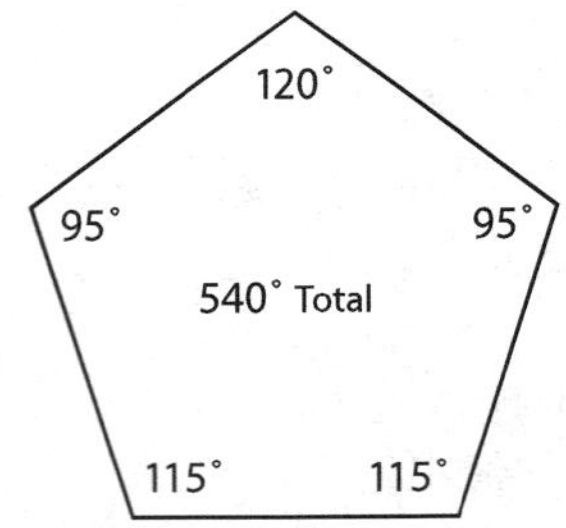

Steps	**Example**
1. Count the angles	For the pentagon: 5
2. Subtract 2	$5 - 2 = 3$
3. Multiply by 180°	$3 \times 180 = 540°$

DRILL **POLYGONS**

A dodecagon is a 12-sided polygon. What's the sum of its angles?

Answer

Steps	**Example**
1. Count the angles	12
2. Subtract 2	$12 - 2 = 10$
3. Multiply by 180°	$10 \times 180 = 1800°$

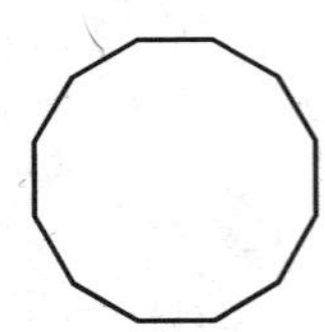

DETAIL **POLYGONS**

If you forget this method, there's another way to find the angle totals.

1. The triangle totals 180°
2. Each additional angle adds 180° to the total.

3 angles

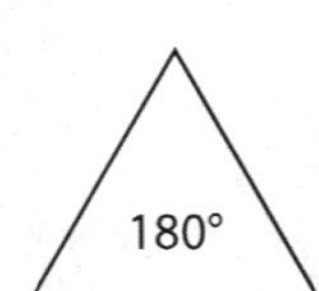

180°

4 angles

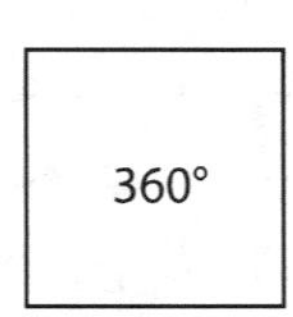

180° + 180°

5 angles

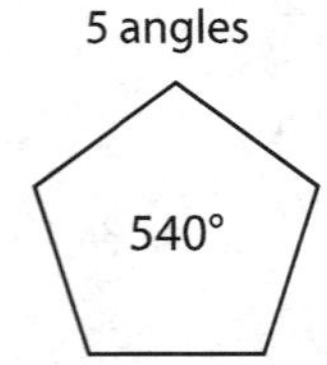

180° + 180° + 180°

TOPIC	FINDING THE THIRD SIDE LENGTH	GEOMETRY
		TRIANGLES

A triangle's side length is limited by the other two side lengths.

LIFELINE 78

To find the possible third side length of a triangle, draw a picture or use the formula.

In this triangle, the unknown side *c* can only be certain lengths.

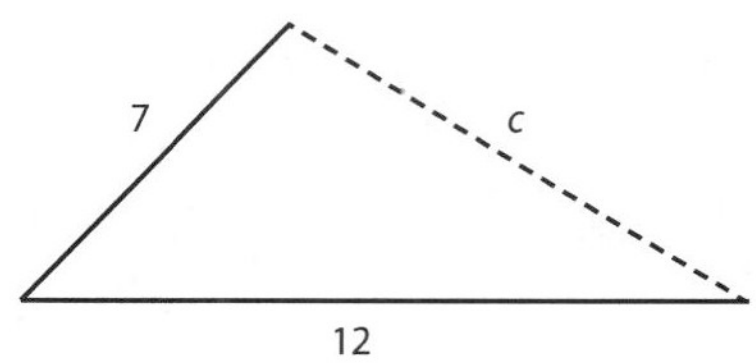

To find the shortest side possible:

For example, it couldn't be 2, because the sides wouldn't reach.

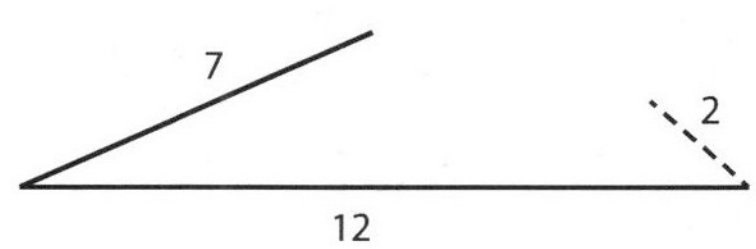

It couldn't be 5, because that would create two flat lines.

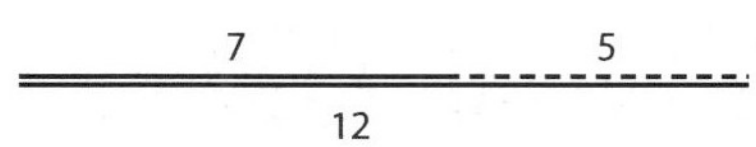

The third side has to be greater than 5, which is 12 – 7.

To find the longest side possible:

For example, it couldn't be 25, because the side would overextend.

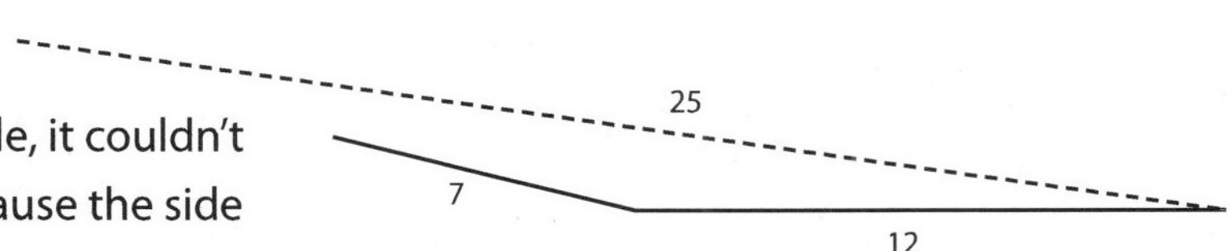

It couldn't be 19 because that would create two flat lines.

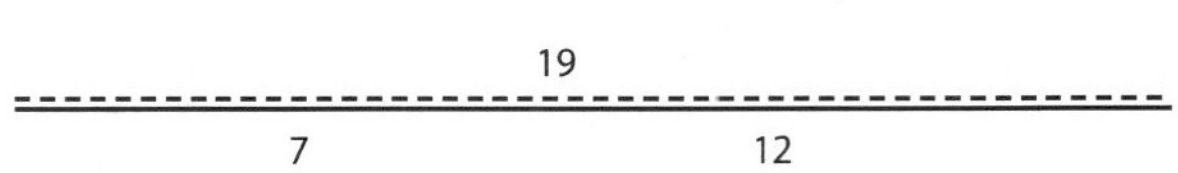

The third side has to be less than 19, which is 12 + 7.

GEOMETRY

TRIANGLES

DETAIL | FINDING THE THIRD SIDE LENGTH

You can also use this formula.

$|a - b| < c < a + b$

Where *a* and *b* represent the side lengths you know, and *c* represents the side length you don't know.

Steps	Example
1. Write down the formula	$\|a - b\| < c < a + b$
2. Using the triangle from the previous page, substitute the side lengths for *a* and *b*	$\|12 - 7\| < c < 12 + 7$
3. Simplify the math to get the possible third side lengths.	$5 < c < 19$

The third side *c* has to be greater than 5, but can't be *equal* to 5.

DETAIL | FINDING THE THIRD SIDE LENGTH

In this formula, the expression $|a - b|$ is in the absolute value, so that it doesn't matter what order you place the sides in.

For example, $|7 - 12|$ equals $|-5|$, which is 5.

For more on the **absolute value**, see Algebra, p. 133.

If 2 and 7 are the lengths of two sides of a triangle, which of the following could be the length of the third side?

I. 5
II. 9
III. 13

(A) II only
(B) I and II
(C) II and III
(D) I, II, and III
(E) None of these

SOLUTION NEXT PAGE

GEOMETRY

TRIANGLES

QUIZ 1 FINDING THE THIRD SIDE LENGTH

If 2 and 7 are the lengths of two sides of a triangle, which of the following could be the length of the third side?

I. 5
II. 9
III. 13

(A) II only
(B) I and II
(C) II and III
(D) I, II, and III
(E) None of these

SOLUTION

STEP 1 Separate the information from the question

2 and 7 are the lengths of two sides of a triangle

STEP 2 Turn this into something you can use

Drawing a picture helps organize your thoughts.

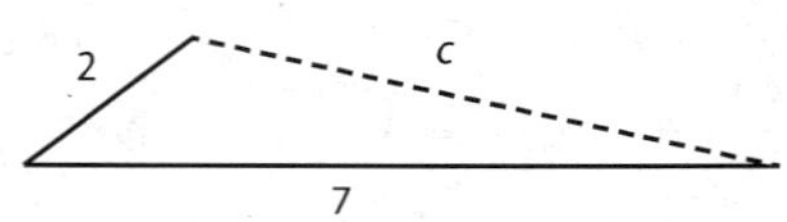

STEP 3 Use this to answer the question

Steps	**Example**
1. Write down the formula	$\lvert a-b \rvert < c < a+b$
2. Substitute the given side lengths for *a* and *b*	$\lvert 2-7 \rvert < c < 2+7$
3. Simplify the math to get the possible third side lengths.	$5 < c < 9$

Because the third side *c* has to be greater than 5 or less than 9, it can't equal 5 or 9, or 13 for that matter.

Answer is E

If two sides of a triangle have lengths of 5, which of the following could be the perimeter of the triangle?

I. 10
II. 15
III. 20

(A) II only
(B) I and II
(C) II and III
(D) I, II, and III
(E) Neither I, II, nor III

The **perimeter** is the sum of all the sides.

SOLUTION NEXT PAGE

GEOMETRY | QUIZ 2 | SIDE LENGTHS

TRIANGLES

If two sides of a triangle have a length of 5, which of the following could be the perimeter of the triangle?

I. 10
II. 15
III. 20

SOLUTION

STEP 1 Separate the information from the question

Two sides of a triangle have a length of 5

STEP 2 Turn this into something you can use

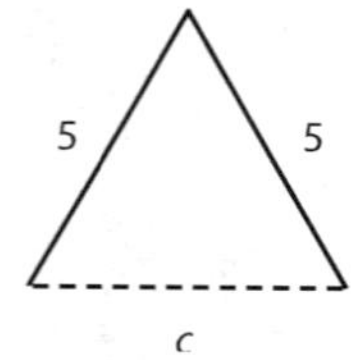

STEP 3 Use this to answer the question

Steps	**Example**
1. Write down the formula	$\|a - b\| < c < a + b$
2. Substitute the given side lengths for *a* and *b*	$\|5 - 5\| < c < 5 + 5$
3. Simplify the math to get the possible third side lengths.	$0 < c < 10$

The third side *c* has to be greater than 0 or less than 10.

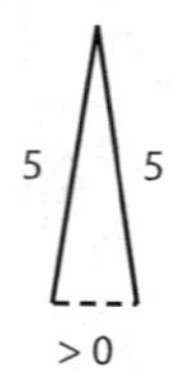

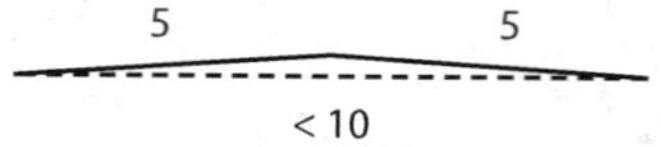

This means that the perimeter has to be greater than 10 or less than 20.

Answer is A

TOPIC **AREA**

GEOMETRY

TRIANGLES

A rectangle with a base of 4 and a height of 6 has an area of 24.

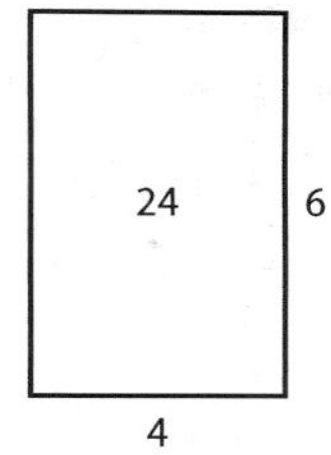

Divide the rectangle in half, corner to corner. Each resulting triangle has an area of 12.

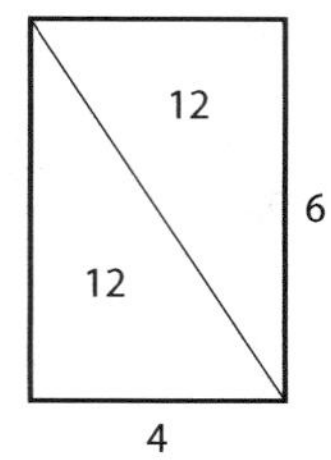

Of the three triangle concepts, area is the least commonly asked on the GMAT.

Base is the length of one side of the triangle.

Height is the distance from the top point of the triangle to the line that the base is on. It's represented by a dotted line with a right-angle box.

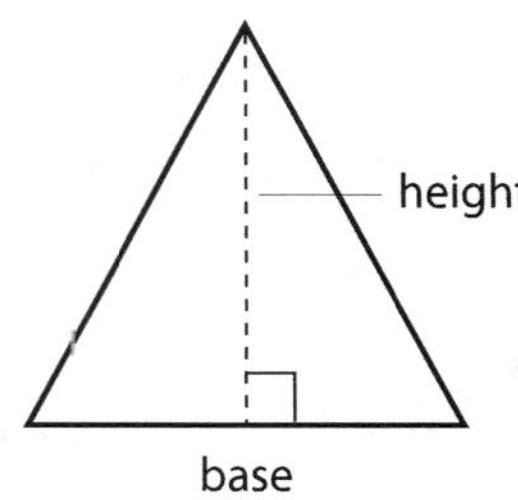

The line is dotted because it's not actually part of the triangle.

The right-angle box tells you that the dotted line is the shortest line possible from the point to the base.

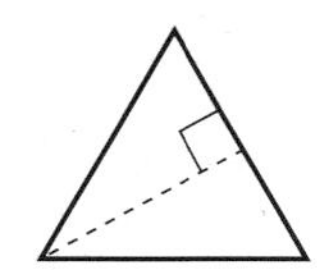

The base isn't always on the bottom.

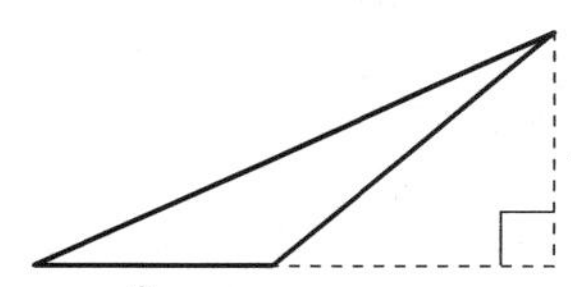

The height can be outside the triangle. This triangle has an area of 6.

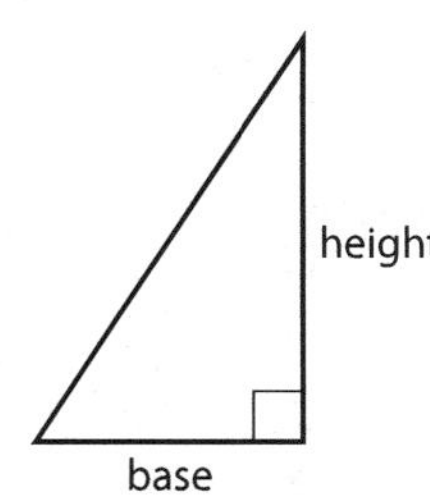

With a right triangle, the base and height are the two sides adjacent to the right angle.

LIFELINE 79

For the area of a triangle, find the base and the height.

$$\text{area} = \frac{\text{base} \times \text{height}}{2} \text{ or } \frac{bh}{2}$$

QUIZ **AREA**

In the rectangular coordinate system shown, what is the area of triangle *PQR* ?

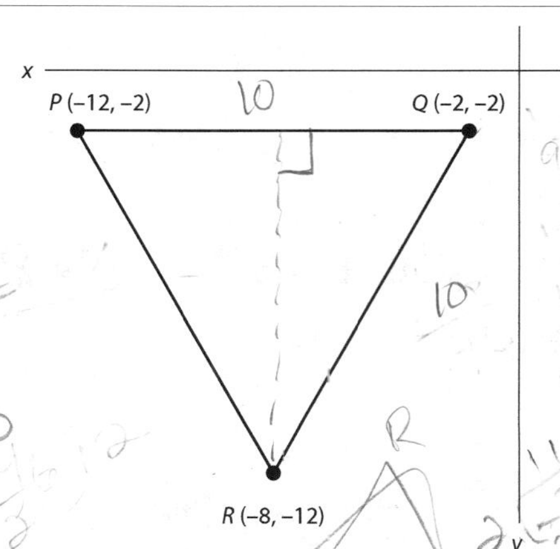

(A) 24

(B) 48

(C) 50

(D) 100

(E) 120

SOLUTION NEXT PAGE

GEOMETRY | QUIZ | AREA

TRIANGLES

In the rectangular coordinate system shown, what is the area of triangle *PQR* ?

(A) 24
(B) 48
(C) 50
(D) 100
(E) 120

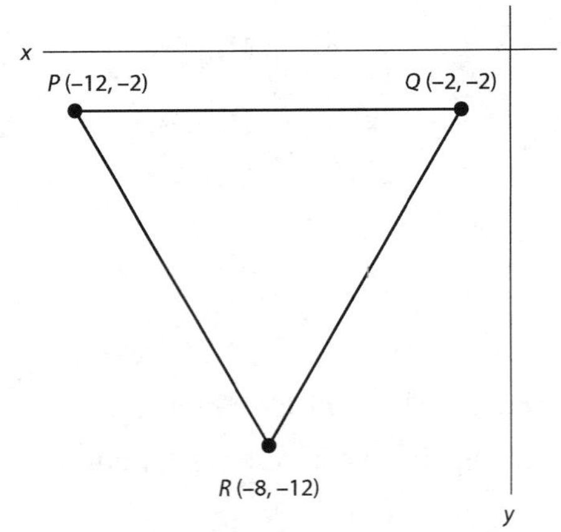

SOLUTION

STEP 1 Separate the information from the question

Copy the drawing down, without the *xy* grid.

STEP 2 Turn this into something you can use

Using the coordinate points, measure the base and height and add these to the drawing.

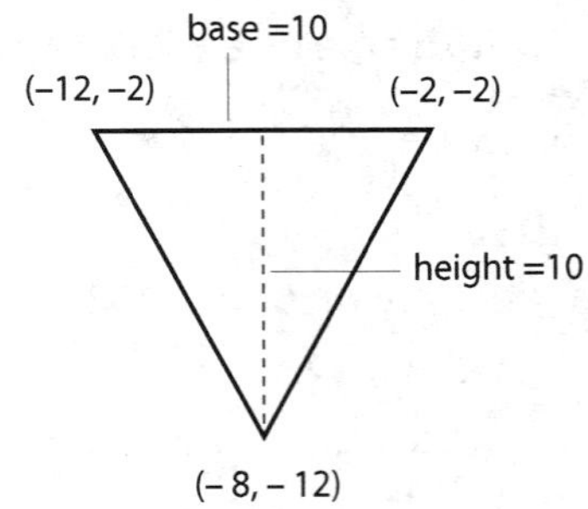

STEP 3 Use this to answer the question

Using the formula $\text{area} = \dfrac{\text{base} \times \text{height}}{2}$

Substitute the base and height of the triangle $= \dfrac{10 \times 10}{2}$

$= 50$

Answer is C

thinktankSCHOLASTIC

TOPIC | **ANGLE TOTALS** | **GEOMETRY**

TRIANGLES

The angles of a triangle always total 180°.

LIFELINE 80

When you know two angles of a triangle, find the third angle.

When there are numbers, find a numeric answer.

Triangle angles total 180°, making $x = 50°$

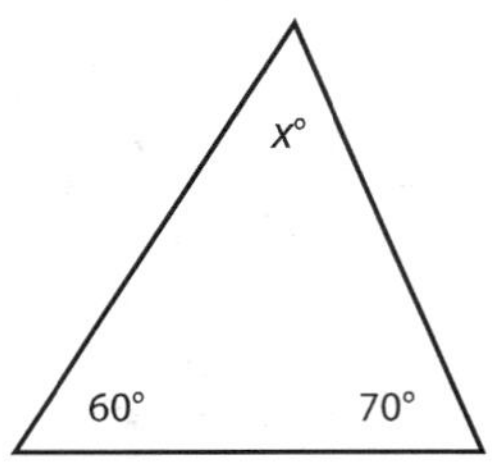

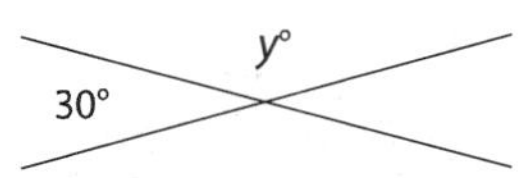

Supplementary angles total 180°, making $y = 150°$

LIFELINE 81

When the angles are given as letters, use those letters to represent the unknown angle.

When there are letters, find the unmarked angle in terms of the letters.

The unmarked angle can be represented by $180 - (x + y)°$

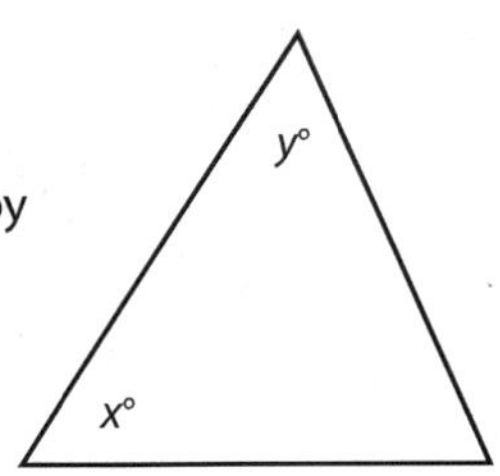

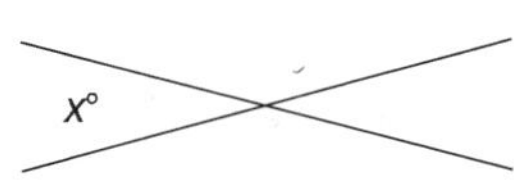

The unmarked angle can be represented by $(180 - x)°$

QUIZ 1 — ANGLE TOTALS

What is the value of z in the figure shown?

(A) 100
(B) 110
(C) 115
(D) 120
(E) 130

SOLUTION NEXT PAGE

GEOMETRY | **QUIZ 1** | **ANGLE TOTALS**

TRIANGLES

What is the value of *z* in the figure shown?

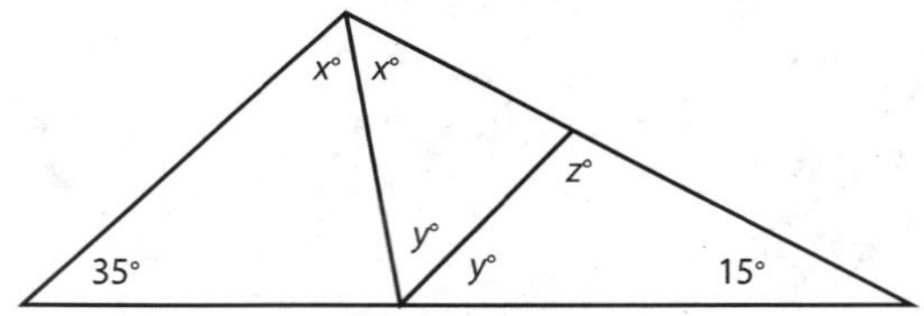

SOLUTION

STEP 1 | **Separate the information from the question**

Copy the drawing down.

STEP 2 | **Turn this into something you can use**

Work the angles separately, starting with the large triangle.

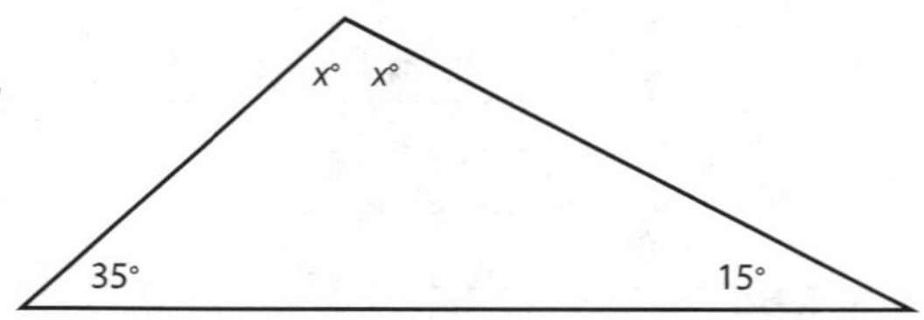

The missing angle is 130°, making $2x = 130$ and $x = 65$

STEP 3 | **Use this to answer the question**

Replace the *x*'s with 65's.

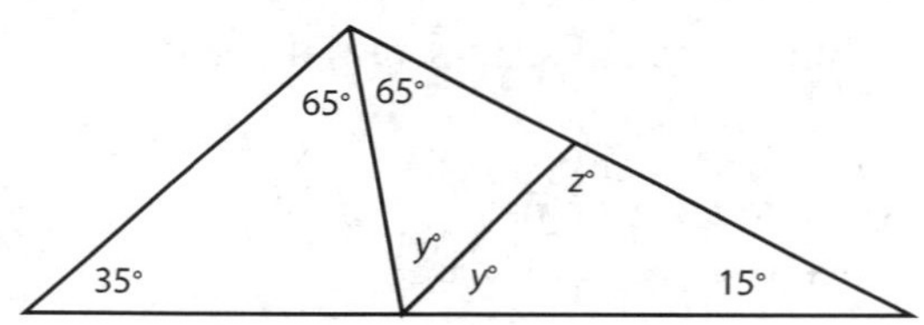

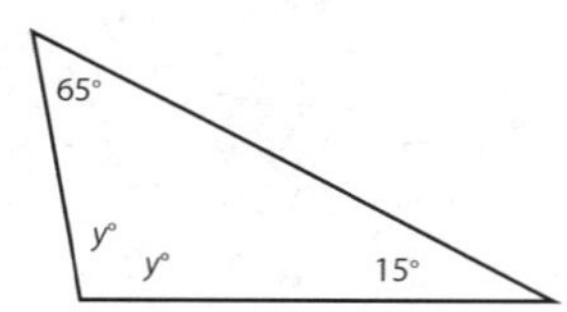

Now solve for *y*.

The missing angle is 100°, making $2y = 100$ and $y = 50$

Replace the *y*'s with 50's.

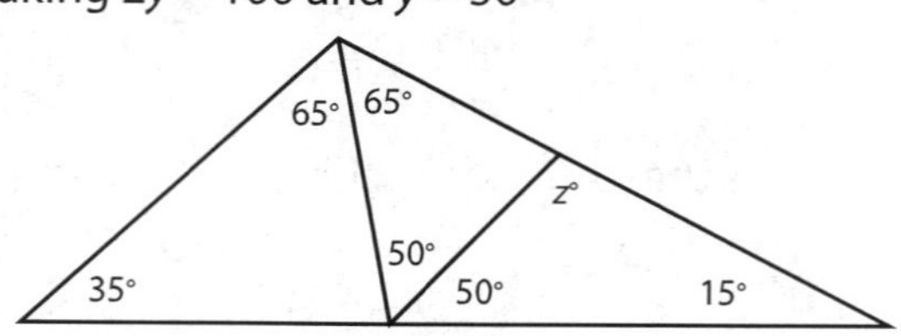

Finally, work the last small triangle with the *z*.

The missing angle *z* is 115°

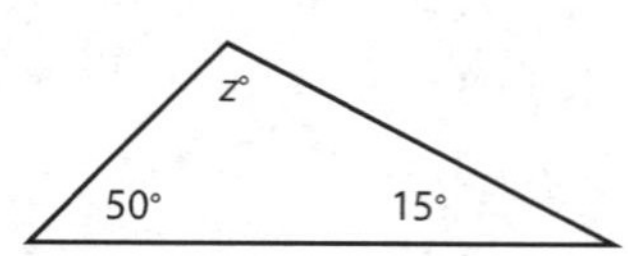

Answer is C

In the triangle shown, y is equivalent to which of the following?

(A) x
(B) $2x$
(C) $3x$
(D) $\frac{x}{2}$
(E) $\frac{2x}{3}$

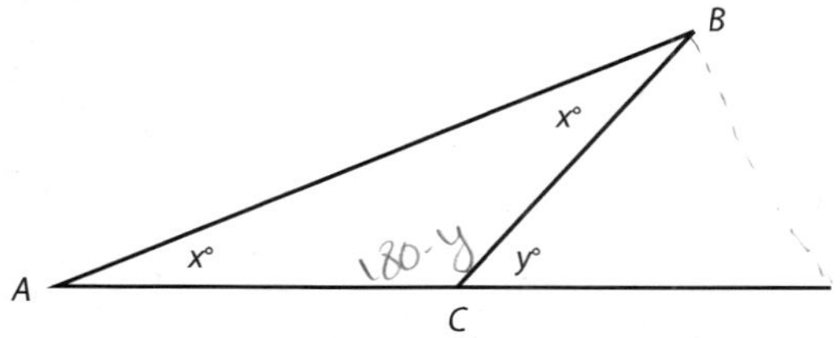

SOLUTION NEXT PAGE

GEOMETRY | QUIZ 2 | ANGLE TOTALS

TRIANGLES

In the triangle shown, y is equivalent to which of the following?

(A) x
(B) $2x$
(C) $3x$
(D) $\frac{x}{2}$
(E) $\frac{2x}{3}$

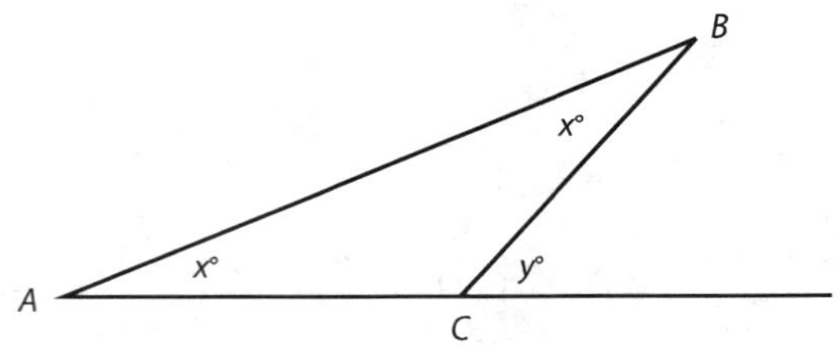

SOLUTION

STEP 1 Separate the information from the question

Copy down the drawing

STEP 2 Turn this into something you can use

Because supplemental angles total 180°,
the angle opposite y equals $(180 - y)°$

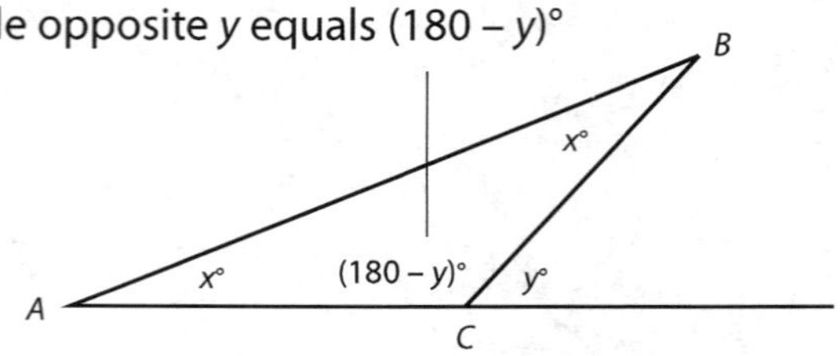

Because the triangle totals 180°,
the third angle equals $(180 - 2x)°$

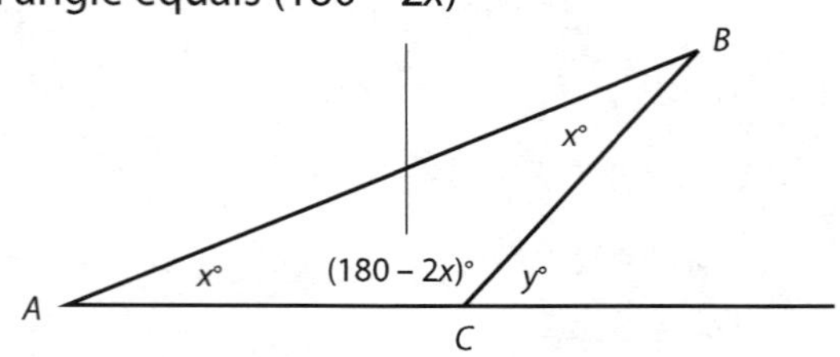

STEP 3 Use this to answer the question

Because the angle equals both $(180 - y)°$ and $(180 - 2x)°$,

$$(180 - y)° = (180 - 2x)°$$
$$180 - y = 180 - 2x$$
$$-y = -2x$$
$$y = 2x$$

Answer is B

DETAIL — THE EQUILATERAL TRIANGLE

An **equilateral** triangle has two features.

1) Each angle is 60°
2) Each side is the same length

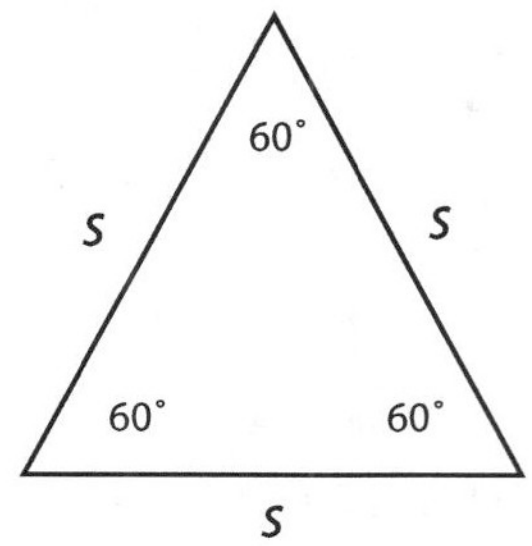

A triangle that's equilateral is also isosceles, like a square is a rectangle.

DETAIL — THE ISOSCELES TRIANGLE

An **isosceles** triangle has three features.

1) Two angles are the same
2) Two sides are the same
3) It's symmetrical

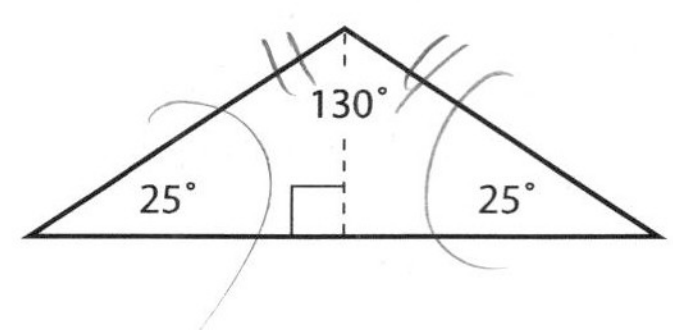

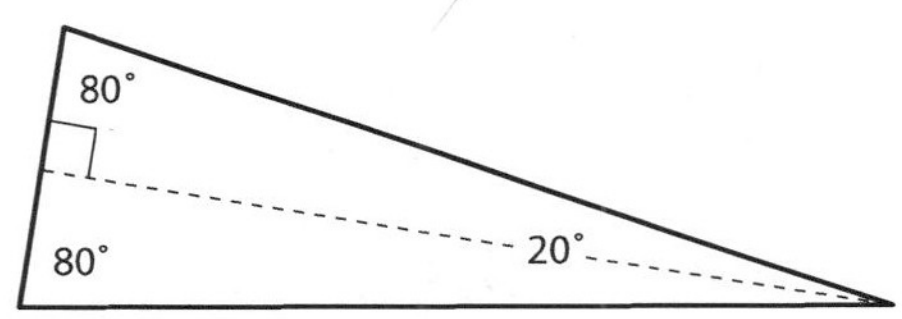

DETAIL — THE RIGHT TRIANGLE

A right triangle has a single 90° angle.

Given any two sides of the right triangle, the third side can be found using the Pythagorean Theorem:

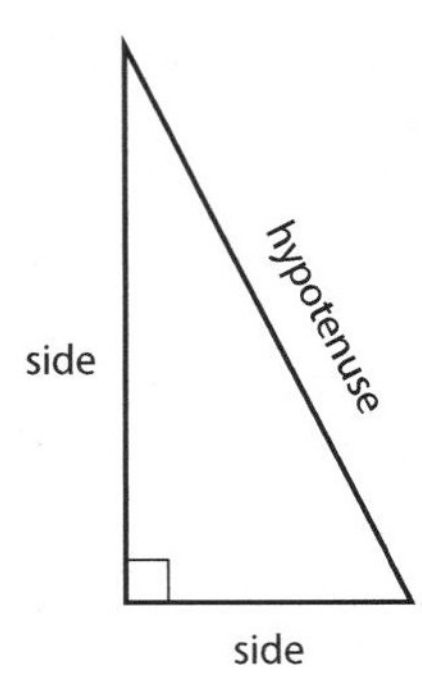

$$a^2 + b^2 = c^2$$

The ***c*** in the theorem represents the **hypotenuse** of the triangle, which is both opposite the right angle and the longest side of the triangle.

Only a right triangle has a hypotenuse.

SOLUTION NEXT PAGE

GEOMETRY | **DETAIL** | **PYTHAGOREAN THEOREM**

TRIANGLES

When a right triangle is missing a side length, you can find it from the other two side lengths.

What is the height of this right triangle?

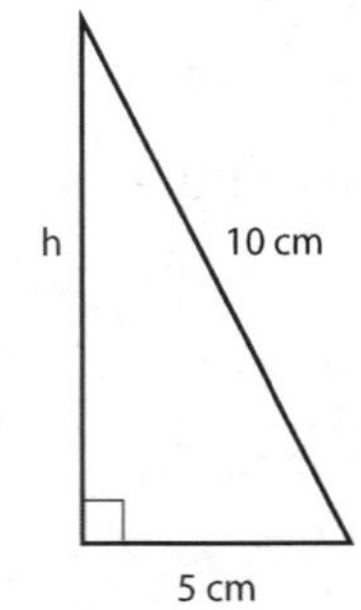

Steps	**Example**
1. Copy the Pythagorean Theorem	$a^2 + b^2 = c^2$
2. Add the sides that you know	$5^2 + b^2 = 10^2$
3. Solve for the missing side	$25 + b^2 = 100$
	$b^2 = 75$
	$b = \sqrt{75}$
	$b = \sqrt{25} \times \sqrt{3}$
	$b = 5\sqrt{3}$

Leave the answer in terms of the radical. For more on this see Arithmetic, page 99.

What is the base of this right triangle?

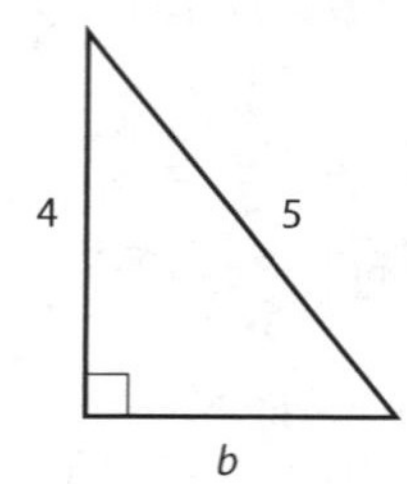

Steps	**Example**
1. Copy the Pythagorean Theorem	$a^2 + b^2 = c^2$
2. Add the sides that you know	$4^2 + b^2 = 5^2$
3. Solve for the missing side	$16 + b^2 = 25$
	$b^2 = 9$
	$b = 3$

This is a common right triangle known as the 3 : 4 : 5 triangle. More on this on the next page.

DETAIL — COMMON RIGHT TRIANGLES

The 3 : 4 : 5 and 5 : 12 : 13 triangles fit easily into the Pythagorean Theorem, with side lengths that are whole numbers. Recognizing these triangles lets you skip the Pythagorean Theorem.

The 3 : 4 : 5 triangle.

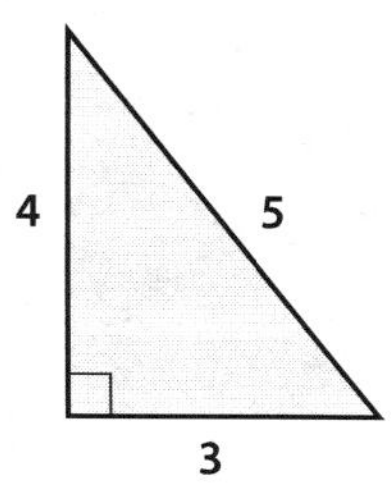

The 5 : 12 : 13 triangle
(Less common)

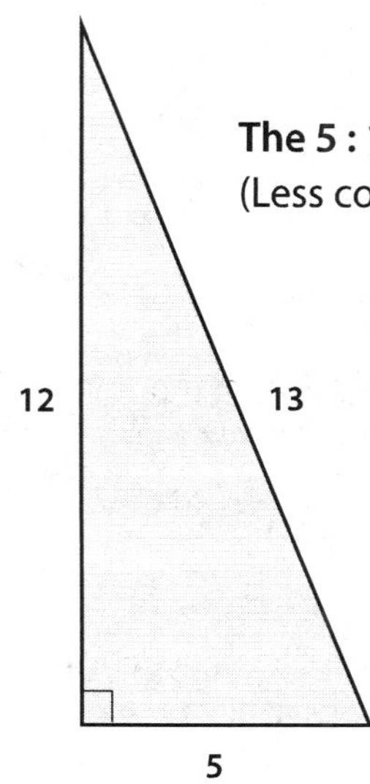

The side lengths are ratios. For example:
The triangle on the lower right is in the ratio of 3 : 4 : 5.

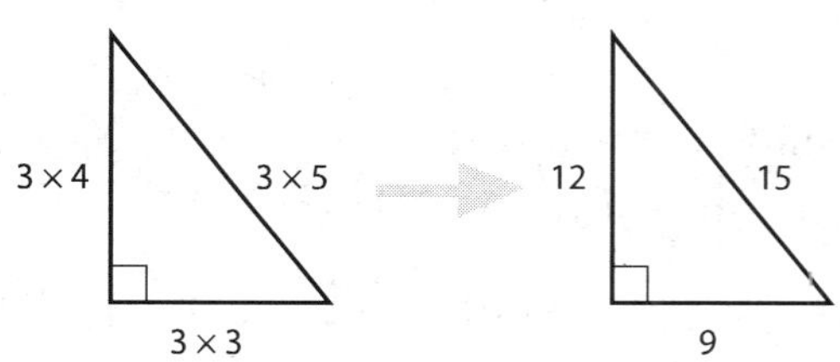

QUIZ — COMMON RIGHT TRIANGLES

In the figure shown,
what is the value of x ?

(A) 60
(B) 65
(C) 70
(D) 75
(E) 80

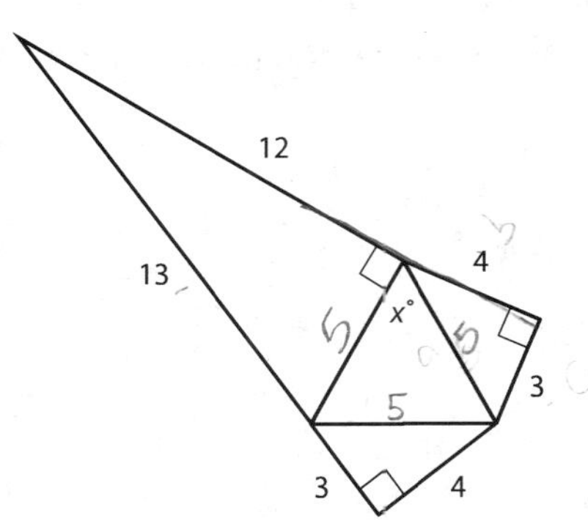

SOLUTION NEXT PAGE

GEOMETRY | QUIZ | COMMON RIGHT TRIANGLES

TRIANGLES

In the figure shown, what is the value of x ?

(A) 60
(B) 65
(C) 70
(D) 75
(E) 80

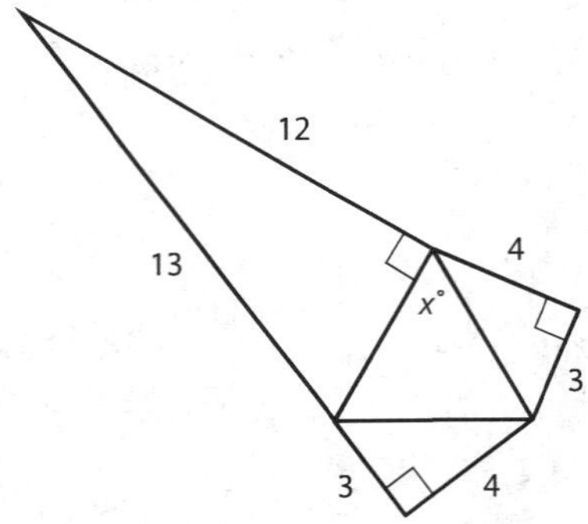

SOLUTION

STEP 1 Separate the information from the question

Copy down the drawing.

STEP 2 Turn this into something you can use

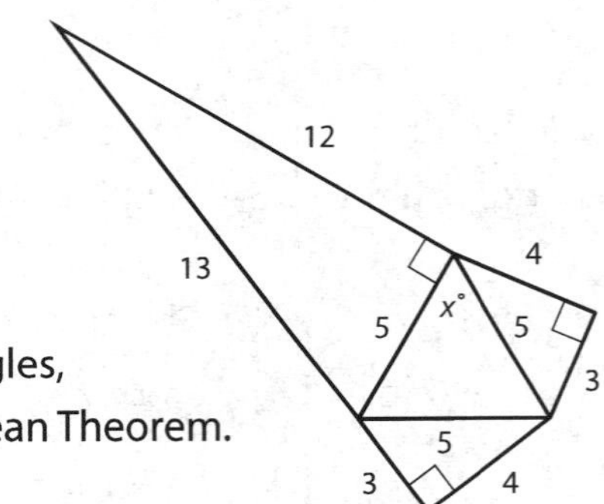

Knowing the common right triangles, there is no need for the Pythagorean Theorem.

These are 3 : 4 : 5 and 5 : 12 : 13 triangles, so add the side lengths of 5 to the drawing.

STEP 3 Use this to answer the question

A triangle with three equal side lengths, in this case the center triangle, is equilateral, making $x = 60$.

Answer is A

 thinktankSCHOLASTIC

TOPIC | **THE 30-60-90 TRIANGLE**

GEOMETRY

TRIANGLES

The 30-60-90 triangle begins as an equilateral triangle, with equal sides and angles.

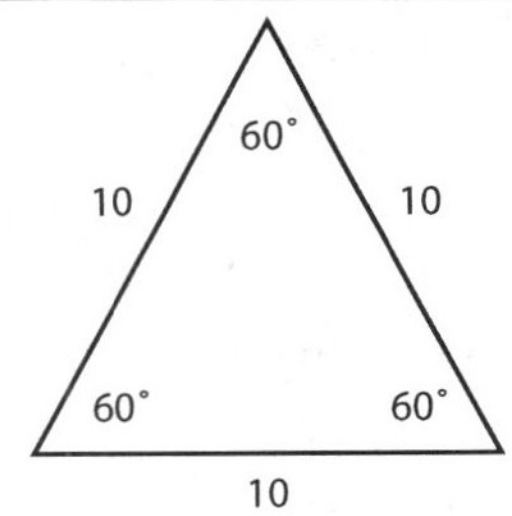

Cutting the triangle in half, the top 60° angle becomes two 30° angles, and the base of 10 becomes two bases of 5.

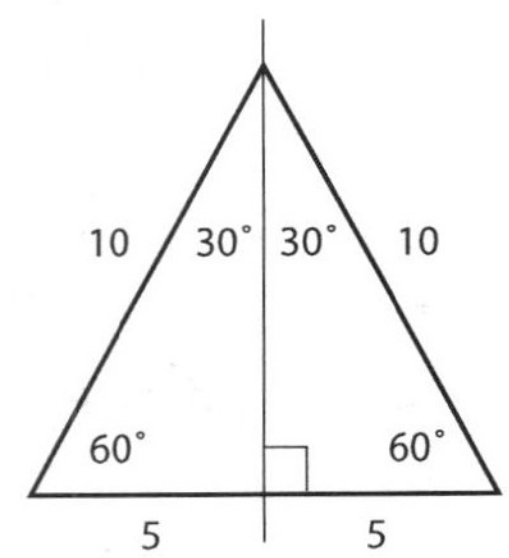

Each resulting triangle is a 30-60-90 triangle with the side ratio $1 : 2 : \sqrt{3}$.

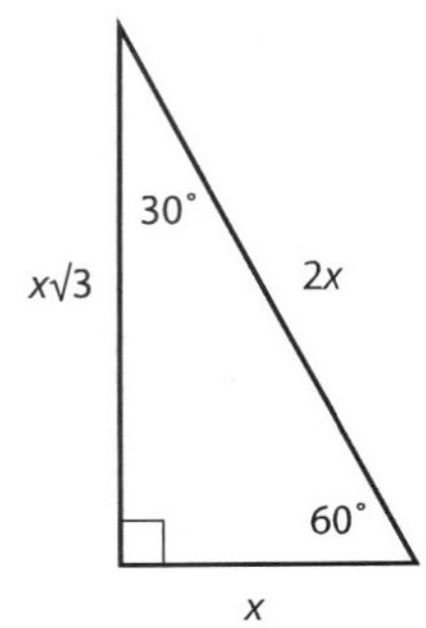

LIFELINE 82

When the base is half the hypotenuse, think of the mnemonic 1-2-3.

Remember
Only a right triangle has a hypotenuse.

1-2-3 refers to the side ratio of $1 : 2 : \sqrt{3}$

QUIZ | **THE 30-60-90 TRIANGLE**

In the *xy* coordinate plane shown, what is the value of point *t* ?

(A) 1
(B) 1.5
(C) $\sqrt{3}$
(D) 2
(E) $2\sqrt{3}$

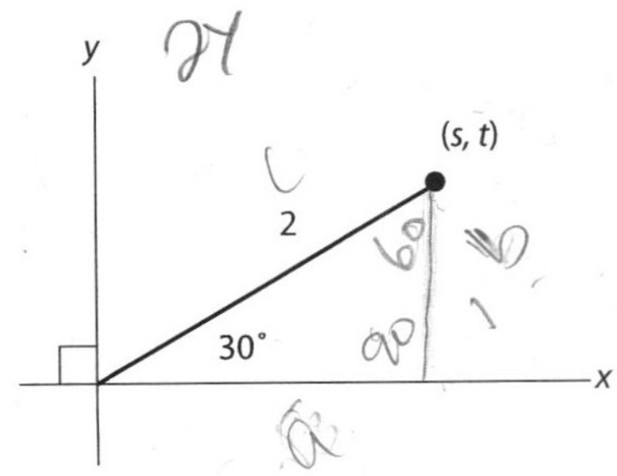

SOLUTION NEXT PAGE

GEOMETRY

TRIANGLES

QUIZ

THE 30-60-90 TRIANGLE

In the *xy* coordinate plane shown, what is the value of point *t* ?

(A) 1
(B) 1.5
(C) $\sqrt{3}$
(D) 2
(E) $2\sqrt{3}$

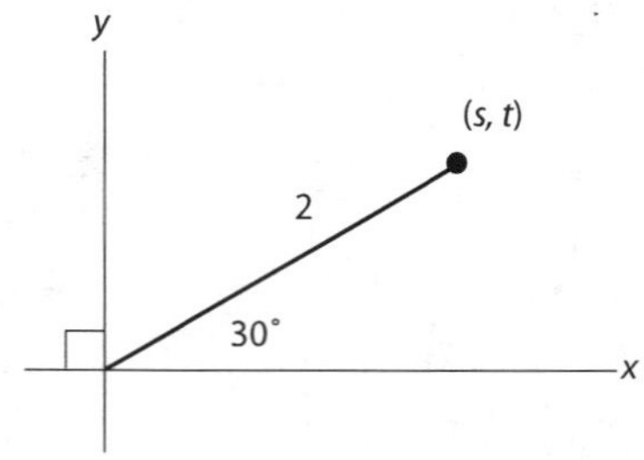

SOLUTION

STEP 1 Separate the information from the question

Copy down the drawing

STEP 2 Turn this into something you can use

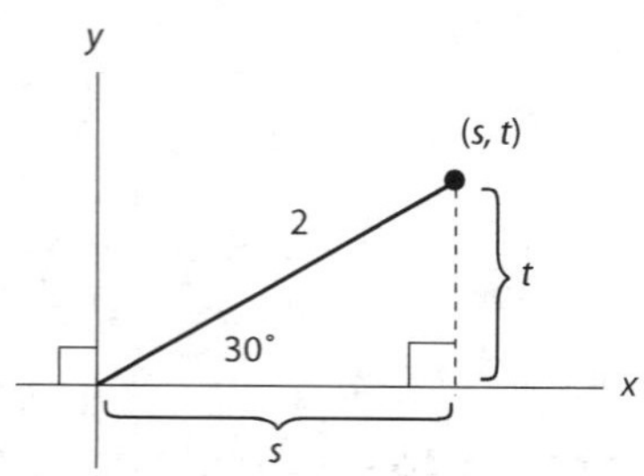

Remember
The *xy* axes of the coordinate plane are perpendicular.

s represents the distance the point traveled "over"

t represents the distance the point traveled "up"

STEP 3 Use this to answer the question

You now have a 30-60-90 triangle with a hypotenuse of 2.

This makes the longer leg (*s*) $\sqrt{3}$ and the shorter leg (*t*) 1.

Answer is A

TOPIC THE 45-45-90 TRIANGLE

The 45-45-90 triangle is from a square cut in half, corner to corner. This triangle is also called an **isosceles right triangle**.

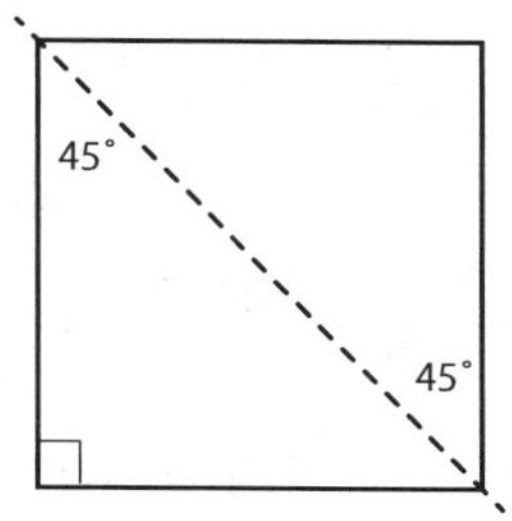

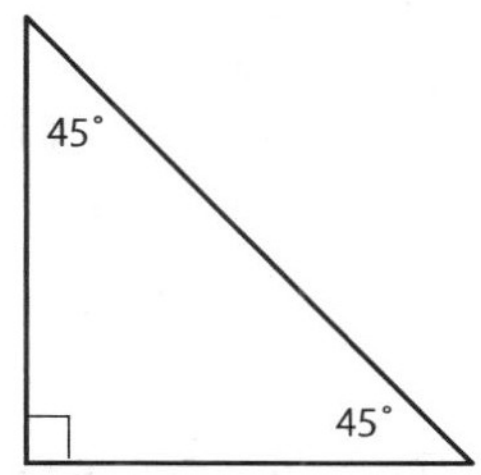

A 45°- 45°- 90° triangle

LIFELINE 83

With the 45-45-90 triangle, there are two side-length ratios, and one of them answers the question.

The ratio that you use depends on whether you have the **side** or the **hypotenuse**.

1) Given the **side**

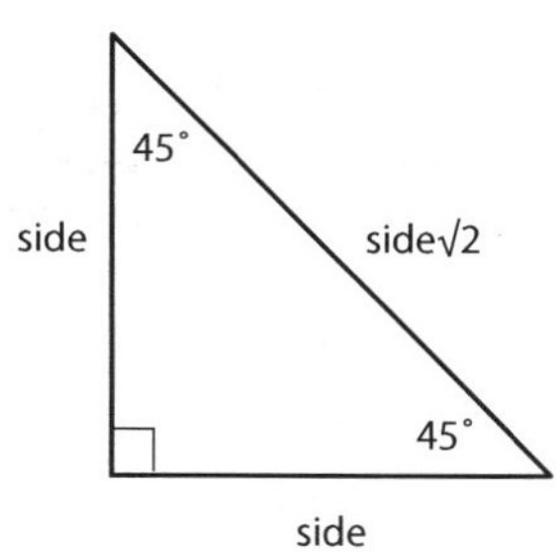

The hypotenuse is the side times $\sqrt{2}$

2) Given the **hypotenuse**

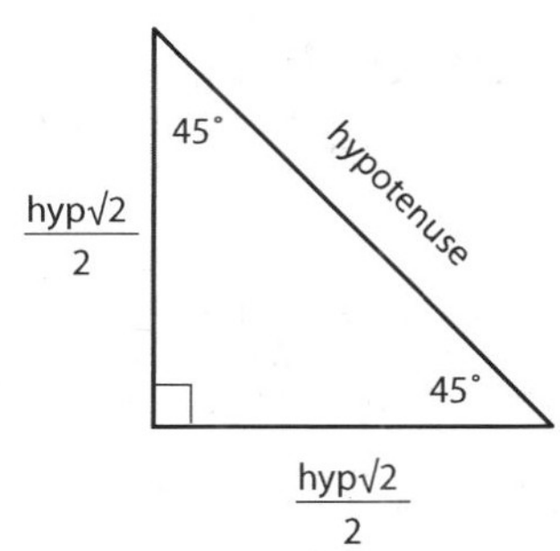

Each side is the hypotenuse times $\sqrt{2}$ over 2

The **side** ratio can be memorized with the mnemonic "1-2" for $1 : \sqrt{2}$.

The **hypotenuse** ratio is more common, but there isn't a mnemonic. You just have to memorize it.

QUIZ 1 THE 45-45-90 TRIANGLE

If a square tile has a 16-inch diagonal, what is the perimeter of the tile?

(A) $16 + 16\sqrt{2}$
(B) $32\sqrt{2}$
(C) 48
(D) 64
(E) 80

SOLUTION NEXT PAGE

GEOMETRY | QUIZ 1 | THE 45-45-90 TRIANGLE

TRIANGLES

If a square tile has a 16-inch diagonal, what is the perimeter of the tile?

(A) $16 + 16\sqrt{2}$
(B) $32\sqrt{2}$
(C) 48
(D) 64
(E) 80

SOLUTION

STEP 1 | **Separate the information from the question**

a square tile has a 16-inch diagonal

STEP 2 | **Turn this into something you can use**

Draw the picture:

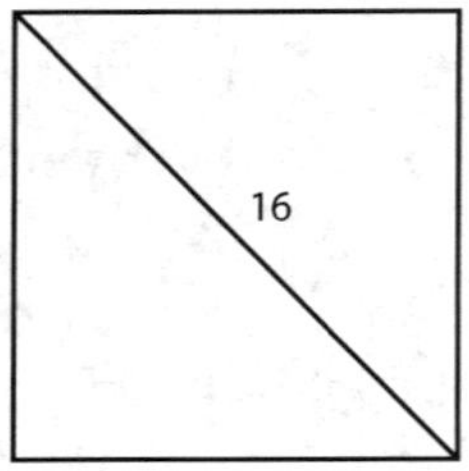

STEP 3 | **Use this to answer the question**

Because the square cut diagonally results in two 45-45-90 triangles, use the side ratios to find each side of the square.

The perimeter (distance around) the square is:

$$8\sqrt{2} + 8\sqrt{2} + 8\sqrt{2} + 8\sqrt{2}$$

or

$$4(8\sqrt{2}) = 32\sqrt{2}$$

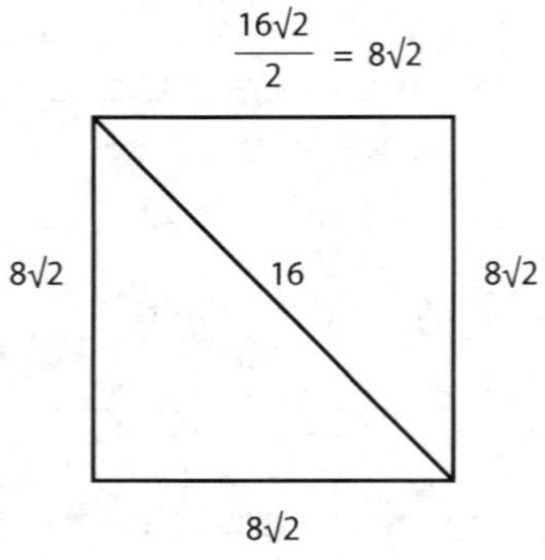

Answer is B

The perimeter of a certain isosceles right triangle is $4 + 4\sqrt{2}$. What is the length of the hypotenuse of the triangle?

(A) 2
(B) $2\sqrt{2}$
(C) 4
(D) $4\sqrt{2}$
(E) 6

SOLUTION NEXT PAGE

GEOMETRY	QUIZ 2	THE 45-45-90 TRIANGLE

TRIANGLES

The perimeter of a certain isosceles right triangle is $4 + 4\sqrt{2}$. What is the length of the hypotenuse of the triangle?

(A) 2
(B) $2\sqrt{2}$
(C) 4
(D) $4\sqrt{2}$
(E) 6

SOLUTION

STEP 1 Separate the information from the question

The perimeter of a certain isosceles right triangle is $4 + 4\sqrt{2}$.

STEP 2 Turn this into something you can use

Draw the picture:

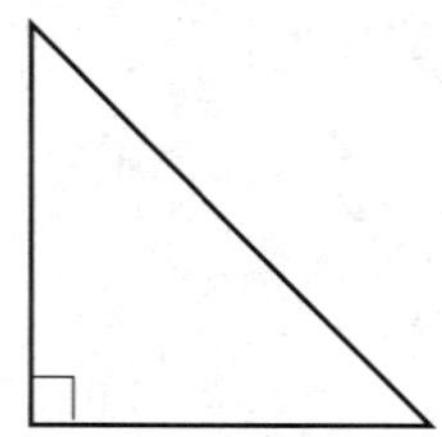

STEP 3 Use this to answer the question

This triangle is one of the two 45-45-90 ratios:

1) If you're given the side

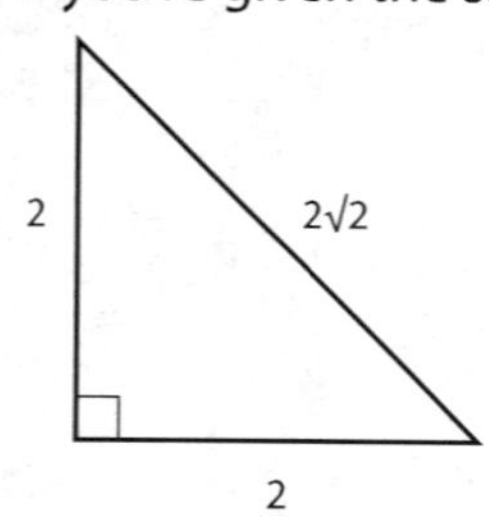

If the side lengths total 4, then each side is 2, making the hypotenuse $2\sqrt{2}$ and the perimeter $4 + 2\sqrt{2}$.

This doesn't match the perimeter in the question, so it isn't this correct triangle.

2) If you're given the hypotenuse

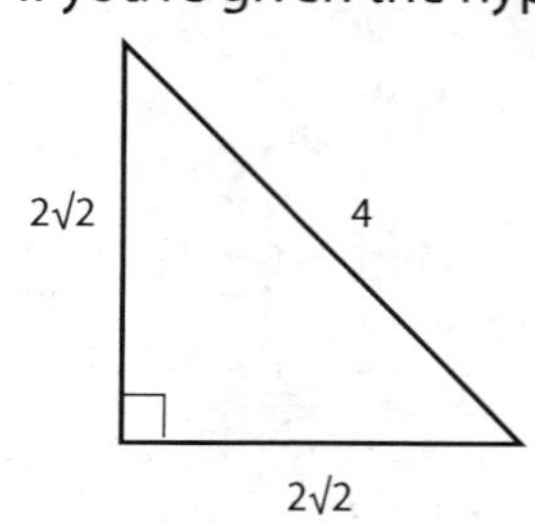

If the hypotenuse is 4, then each side length is $2\sqrt{2}$, making the perimeter of this triangle $4 + 4\sqrt{2}$ which matches the perimeter in the question.

This is the correct triangle, and the hypotenuse is 4.

Answer is C

QUADRILATERALS

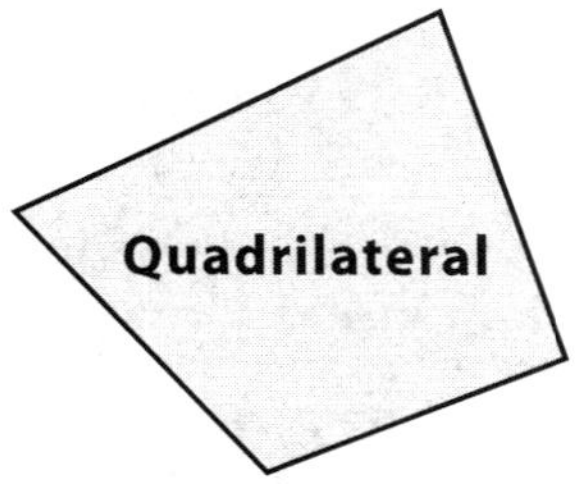

Any shape with four straight sides.

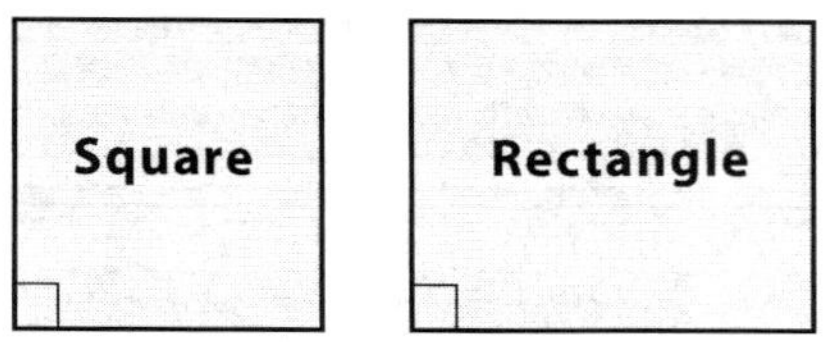

Each angle equals 90°

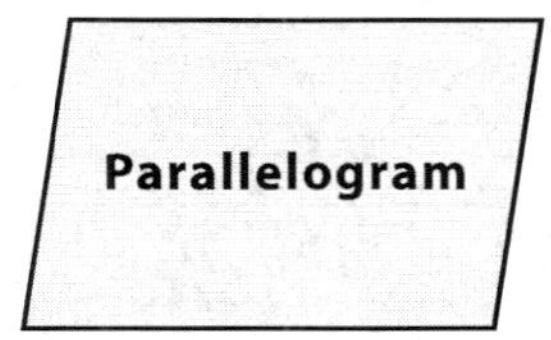

Opposite sides are parallel

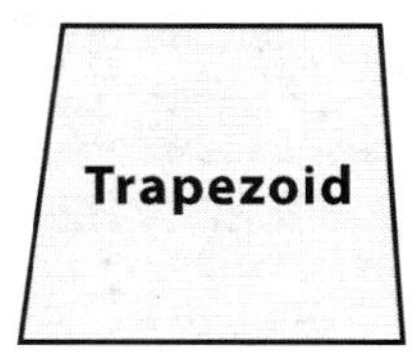

Two sides are parallel

LIFELINE 84

To measure the area of most quadrilaterals, find the base and the height.

The rectangle and the parallelogram.

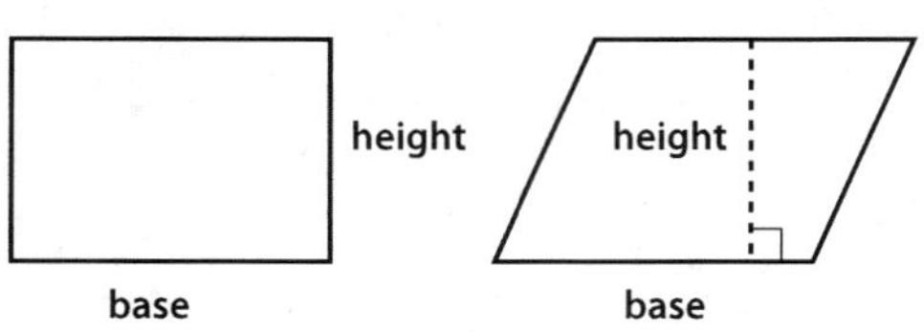

The distance between the two parallel sides is the **height** of the parallelogram.

base × height = area

The square.

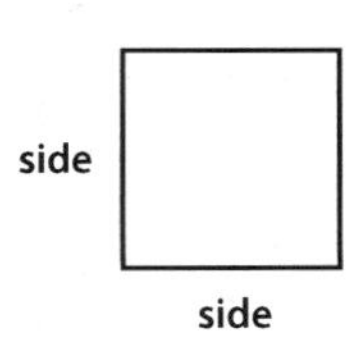

The base and the height of a square are equal, so they're called **sides**.

$(\text{side})^2 = \text{area}$

GEOMETRY | **DETAIL** | **THE TRAPEZOID**

QUADRILATERALS

The area of a trapezoid is also found with base × height.

However, because the trapezoid has two bases, average the bases first. You can also use this equation:

$$\left(\frac{b_1 + b_2}{2}\right)(\text{height})$$

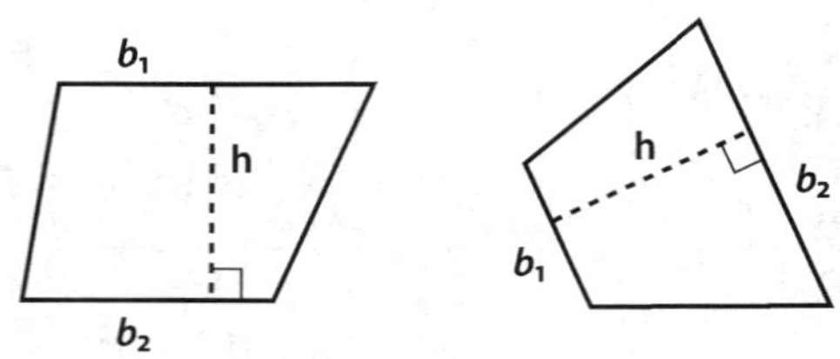

DRILL | **THE TRAPEZOID**

What's the area of this trapezoid?

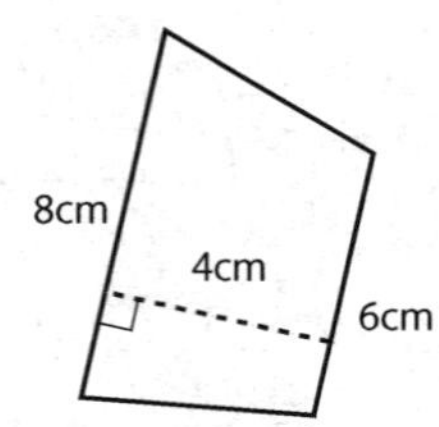

SOLUTION

Drill answer:

$$\left[\frac{6+8}{2}\right]4 = 7 \times 4 = 28$$

TOPIC CIRCLES

A **circle** is a shape with every point on the shape the same distance from the center. This distance is called the **radius**.

The **diameter** is the distance across the circle (twice the radius).
The **circumference** is the distance around the circle.
A **chord** is a line that passes through the circle.
A **tangent** is a line that touches the circle on one point.

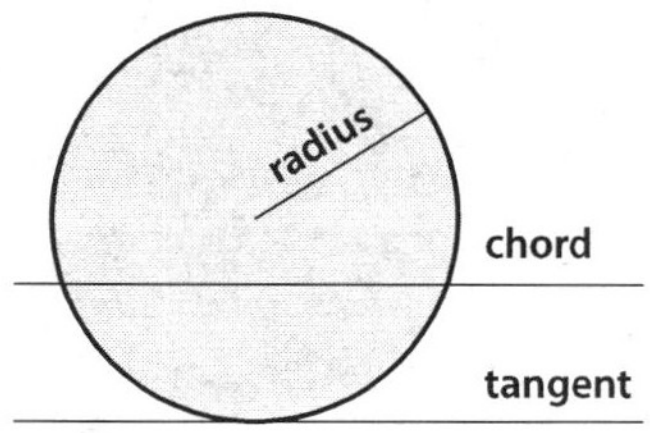

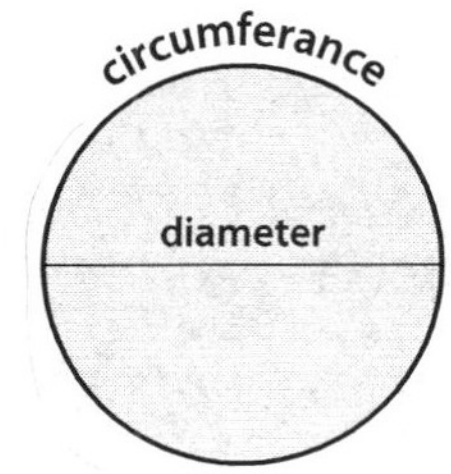

pi is the ratio of the circumference to the diameter

$$\pi = \frac{C}{d} \approx 3.14$$

You don't have to memorize that the value of *pi*. Just know that it's slightly over 3. Answers to circle questions will typically be in terms of *pi*, such as "2π", not "6.28"

LIFELINE 85

When working with circles, leave *pi* as the symbol π , not 3.14

Finding the circumference of a circle:
Because the radius is half of the diameter, either of these formulas works.

$$C = \pi d$$
$$C = 2\pi r$$

Finding the area of a circle:

$$A = \pi r^2$$

DRILL CIRCLES

If a circle has a radius of 6, in terms of *pi*, what are the circumference and area?

SOLUTION NEXT PAGE

GEOMETRY | DRILL | CIRCLES

CIRCLES

SOLUTION

If a circle has a radius of 6, in terms of *pi*, what are the circumference and the area?

Steps	Example
1. Draw the circle	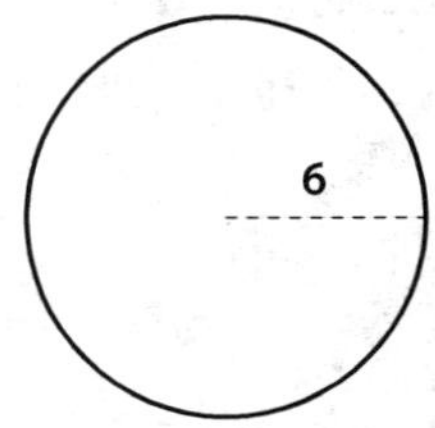
2. Write out the formulas	Circumference $C = 2\pi r$ Area $A = \pi r^2$
3. Plug in the value for *r*	Circumference $C = 2\pi 6$ Area $A = \pi 6^2$
4. Simplify	Circumference $C = 12\pi$ Area $A = 36\pi$

A nylon string 6 feet long is cut into two pieces. One piece is used to form a square with side s, and the other is used to form a circle.

No string is left over. Which of the following represents the total area, in square feet, of the square and the circular regions in terms of s?

A) $\frac{(3-2s)^2}{\pi} + s^2$

B) $\frac{2(3-2s)}{\pi} + s$

C) $(6-4s)\ \pi + s$

D) $\frac{(6-4s)}{\pi} + s^2$

E) $\frac{(6-4s)}{\pi} + s$

SOLUTION NEXT PAGE

GEOMETRY

CIRCLES

QUIZ CIRCLES

A nylon string 6 feet long is cut into two pieces. One piece is used to form a square with side s, and the other is used to form a circle.

No string is left over. Which of the following represents the total area, in square feet, of the square and the circular regions in terms of s?

SOLUTION

STEP 1 Separate the information from the question

A nylon string 6 feet long is cut into two pieces. One piece is used to form a square with side s, and the other is used to form a circle.

STEP 2 Turn this into something you can use

Draw a piece of string cut roughly in half

Draw the square with side lengths s and the circle.

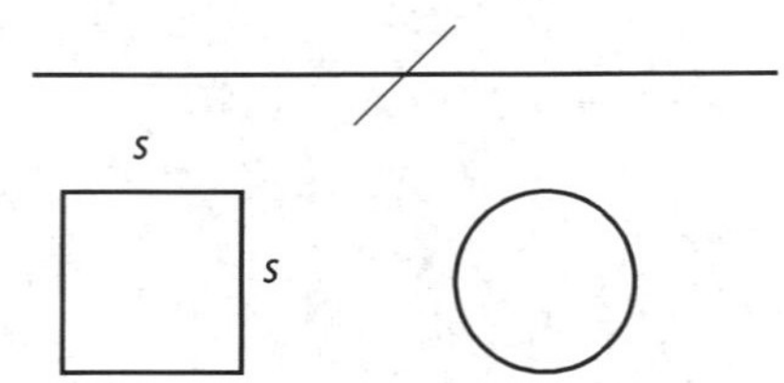

STEP 3 Use this to answer the question

If the square has a side length of s, then $4s$ is used to make the square.

The remaining string, a.k.a. the circumference of the circle, is $6 - 4s$.

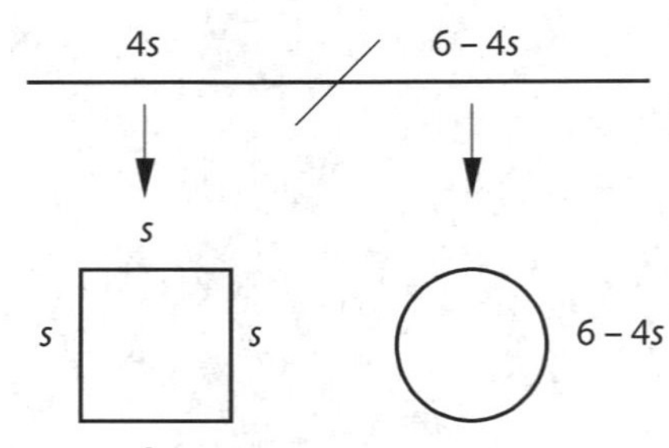

To find the area of the circle, you need the radius.

You have the circumference $(6 - 4s)$, so backsolve that to get the radius.

CONTINUED NEXT PAGE

SOLUTION CONTINUED

To find the radius of the circle

Start with the formula for the circumference $C = 2\pi r$

Substitute (6 – 4*s*) for the circumference $6 - 4s = 2\pi r$

Solve for *r* (the radius) by isolating it $\frac{6 - 4s}{2\pi} = r$

Simplify the fraction $\frac{3 - 2s}{\pi} = r$

To find the area of the circle

Start with the formula for the area $A = \pi r^2$

Plug the value for r $A = \pi \left[\frac{3 - 2s}{\pi}\right]^2$

And simplify.

1. The exponent means the fraction is multiplied by itself $A = \pi \left[\frac{3 - 2s}{\pi}\right]\left[\frac{3 - 2s}{\pi}\right]$

2. Two of the *pi*'s cancel $A = \left[\frac{3 - 2s}{1}\right]\left[\frac{3 - 2s}{\pi}\right]$

3. The two fractions are combined into one $A = \frac{(3 - 2s)^2}{\pi}$

To find the area of the square

Square the side s^2

The total area is the area of the circle plus the area of the square: $\frac{(3 - 2s)^2}{\pi} + s^2$

Answer is A

GEOMETRY

CIRCLES

Though this can be asked with any shape, the GMAT tends to use circles.

Regardless of the shape, the method is the same.

TOPIC SUBTRACTING SHAPES

Some GMAT questions feature two shapes, with one inside the other. The GMAT then asks you for the area of the surrounding (shaded) part as seen here.

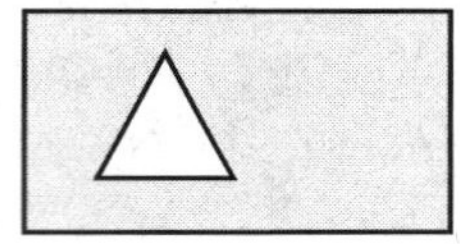

LIFELINE 86

To find the surrounding area of two shapes, subtract the smaller shape from the larger shape.

Steps	**Example**	
1. Draw the two shapes separately		
2. Find the area of each shape	24	6
3. Subtract the smaller shape from the larger shape	24 – 6 = 18	
And the area of the surrounding part is 18.	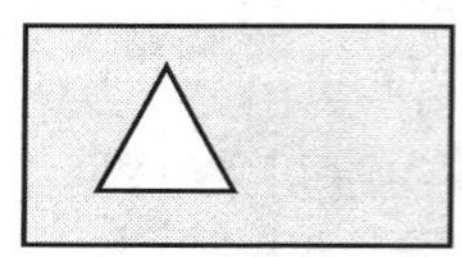	

A circular plate with an 18″ diameter rests on a circular table with a 10″ radius. If no part of the plate is over the edge of the table, what is the visible area of the table?

(A) 2π

(B) 9π

(C) 11π

(D 17π

(E) 19π

SOLUTION NEXT PAGE

GEOMETRY | QUIZ 1 | SUBTRACTING SHAPES

CIRCLES

A circular plate with an 18″ diameter rests on a circular table with a 10″ radius. If no part of the plate is over the edge of the table, what is the visible area of the table?

SOLUTION

STEP 1 Separate the information from the question

A circular plate with an 18″ diameter rests on a circular table with a 10″ radius.

STEP 2 Turn this into something you can use

Draw the picture.

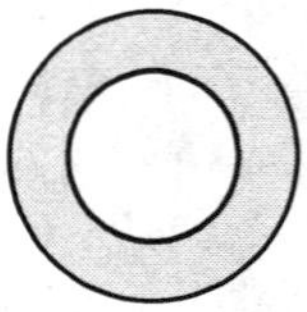

STEP 3 Use this to answer the question

Steps		Example	
1. Draw the two shapes separately		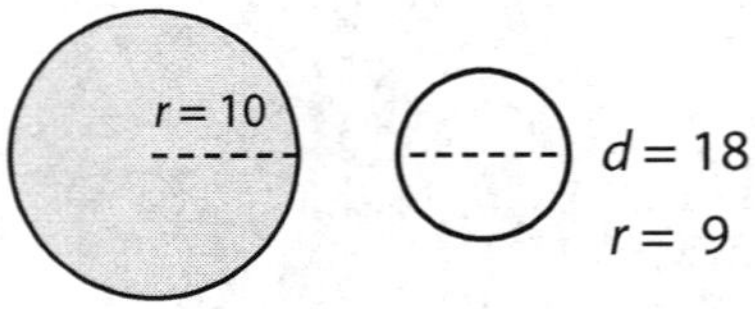	
2. Find the area of each shape	$A = \pi r^2$	100π	81π
3. Subtract the smaller shape from the larger shape		$100\pi - 81\pi = 19\pi$	

Answer is E

 thinktankSCHOLASTIC

In the figure shown, a circle is inscribed within a square.
If the area of the square is 16, what is the area of the shaded region?

(A) 4π
(B) 8π
(C) 16π
(D) $16 - 4\pi$
(E) $16 - 8\pi$

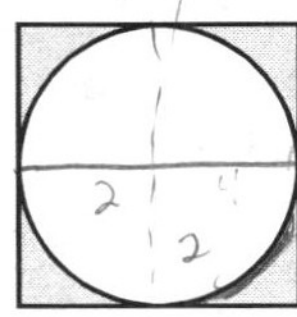

SOLUTION NEXT PAGE

GEOMETRY | QUIZ 2 | SUBTRACTING SHAPES

CIRCLES

In the figure shown, a circle is inscribed within a square.
If the area of the square is 16, what is the area of the shaded region?

(A) 4π
(B) 8π
(C) 16π
(D) $16 - 4\pi$
(E) $16 - 8\pi$

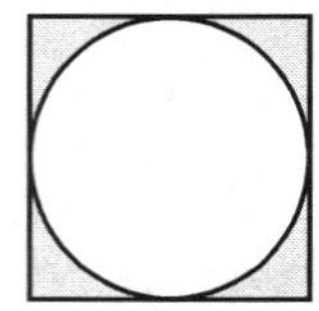

SOLUTION

STEP 1 Separate the information from the question

Copy down the drawing
The area of the square is 16

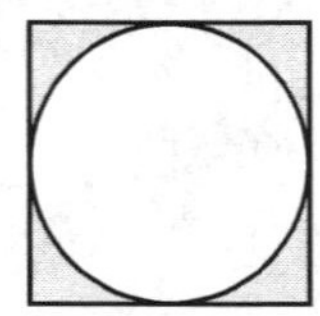

STEP 2 Turn this into something you can use

This means each side of the square is 4, and the diameter of the circle is 4

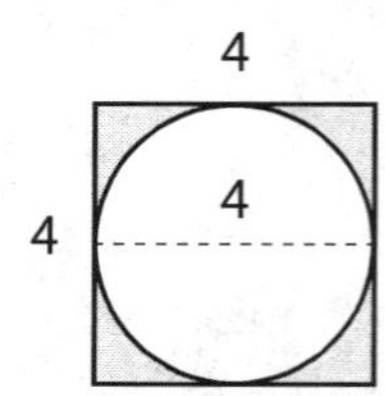

STEP 3 Use this to answer the question

Steps	**Example**	
1. Draw the two shapes separately	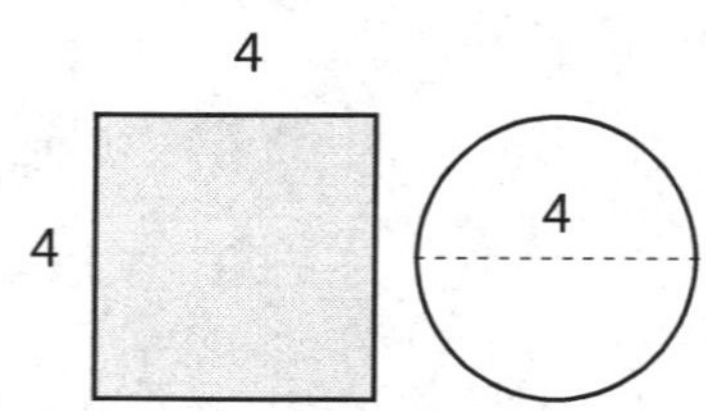	
	s^2	πr^2
2. Find the area of each shape	16	4π
3. Subtract the smaller shape from the larger shape	$16 - 4\pi$	

You don't need to calculate the difference. Just find the pieces, and they'll match an answer choice.

Answer is D

TOPIC | **PARTS OF A CIRCLE**

An **arc** is part of the circle's circumference.
Think of a pizza crust.

A **sector** is part of the circle's area.
Think of a piece of pizza.

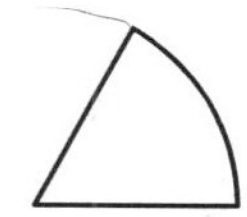

LIFELINE 87

To find the arc or the sector, measure the whole circle and take the fraction.

If you have a circle with a radius of 6:

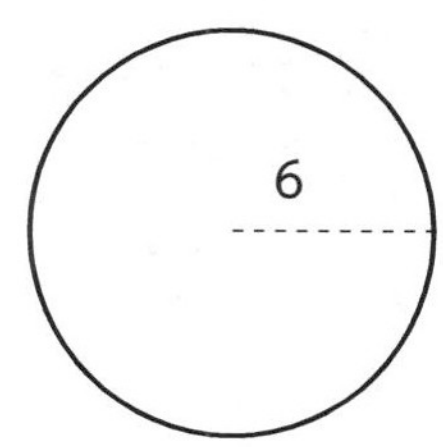

The circumference is 12π and the area is 36π.

Cover half the circle:

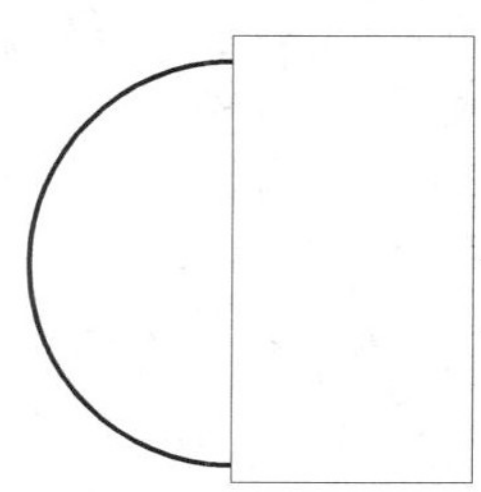

The arc is ½ of the circumference: 6π.
The sector is ½ of the area: 18π.

Cover ¾ of the circle:

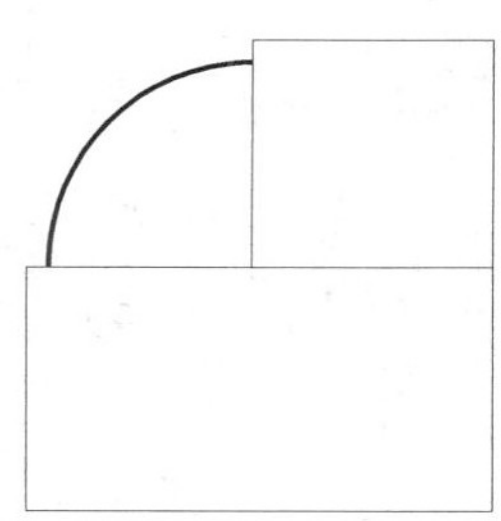

The arc is ¼ of the circumference: 3π.
The sector is ¼ of the area: 9π.

The angle in the center of the circle is 90°, representing ¼ of the circle.

GEOMETRY	DETAIL	PARTS OF A CIRCLE

CIRCLES

Measuring the degree of the angle:

When the angle comes from the **center** of the circle, the degree measure of the arc and the angle are the same.

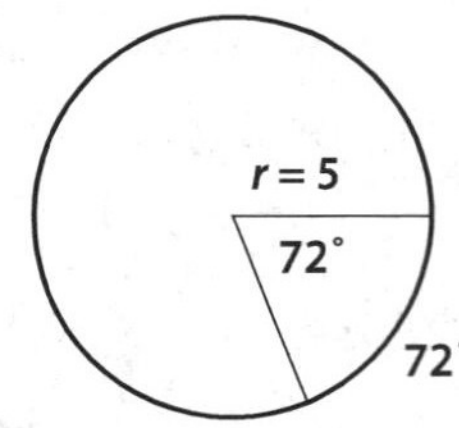

To keep the math simple, two common multiples of arc degree measures are 30° and 36°.

$$\frac{30^\circ}{360^\circ} = \frac{1}{12} \text{ of the circle}$$

$$\frac{36^\circ}{360^\circ} = \frac{1}{10} \text{ of the circle}$$

Measuring the length of the arc:

1. Find the **fraction** of the circle by dividing the angle by 360 — $\frac{72^\circ}{360^\circ} = \frac{1}{5}$
2. Measure the **entire** circumference — Radius is 5; Circumference is 10π
3. Multiply the circumference by the fraction — $10\pi \times \frac{1}{5} = 2\pi$

Measuring the area of the sector:

1. Find the **fraction** of the circle by dividing the angle by 360 — $\frac{72^\circ}{360^\circ} = \frac{1}{5}$
2. Measure the **entire** area — Radius is 5; Area is 25π
3. Multiply the area by the fraction — $25\pi \times \frac{1}{5} = 5\pi$

DRILL	PARTS OF A CIRCLE

If the circle shown has a radius of 6 and an arc of 60°,

1) What's the length of the arc?

2) What's the area of the sector?

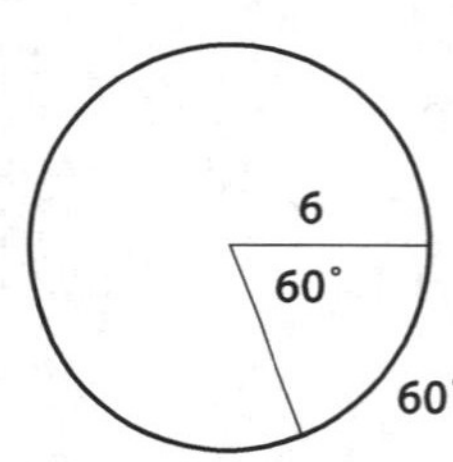

SOLUTION NEXT PAGE

DRILL — PARTS OF A CIRCLE

If the circle shown has a radius of 6 and an arc of 60°,

1) What's the length of the arc?

2) What's the area of the sector?

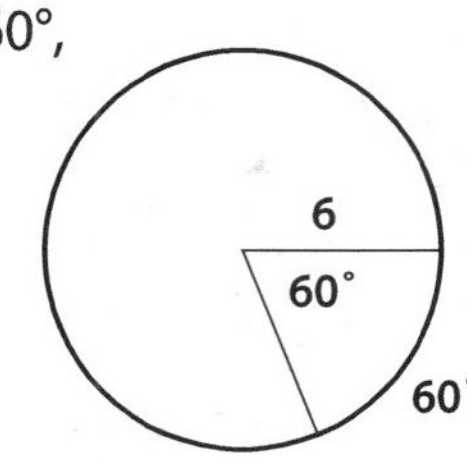

Steps

1. Find the fraction of the circle by dividing the angle by 360 — $\frac{60°}{360°} = \frac{1}{6}$

2. Measure the entire circle — Radius is 6

   To find the arc: use the circumference — $2\pi r = 12\pi$

   To find the sector: use the area — $\pi r^2 = 36\pi$

3. Multiply this by the fraction

   To find the arc: use the circumference — $12\pi \times \frac{1}{6} = 2\pi$

   To find the sector: use the arc — $36\pi \times \frac{1}{6} = 6\pi$

SOLUTION

QUIZ — PARTS OF A CIRCLE

What is the distance, in inches, traveled by the end of the minute hand of a wristwatch in exactly 15 minutes if the minute hand is exactly ½″ long?

(A) π

(B) $\frac{\pi}{2}$

(C) $\frac{\pi}{3}$

(D) $\frac{\pi}{4}$

(E) $\frac{\pi}{5}$

SOLUTION NEXT PAGE

GEOMETRY | QUIZ | PARTS OF A CIRCLE

CIRCLES

What is the distance, in inches, traveled by the end of the minute hand of a wristwatch in exactly 15 minutes if the minute hand is exactly ½" long?

SOLUTION

STEP 1 Separate the information from the question

The ½" minute hand of a wristwatch travels exactly 15 minutes

STEP 2 Turn this into something you can use

Draw the picture:

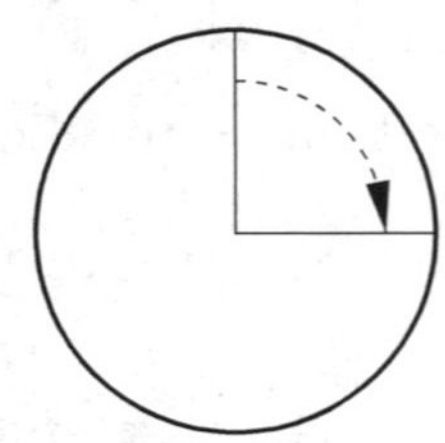

STEP 3 Use this to answer the question

1. Find the fraction of the circle by dividing the angle by 360 $\quad \frac{90°}{360°} = \frac{1}{4}$

2. Measure the entire circle $\quad$ Radius is $\frac{1}{2}$

   You're finding the arc, so measure the entire circumference $\quad 2\pi r = 2\pi\frac{1}{2} = \pi$

3. Multiply this by the fraction $\quad \pi \times \frac{1}{4} = \frac{\pi}{4}$

Answer is D

TOPIC | **ANGLES ON A CIRCLE**

GEOMETRY

CIRCLES

An **angle on a circle** creates an arc with twice the degree measure.

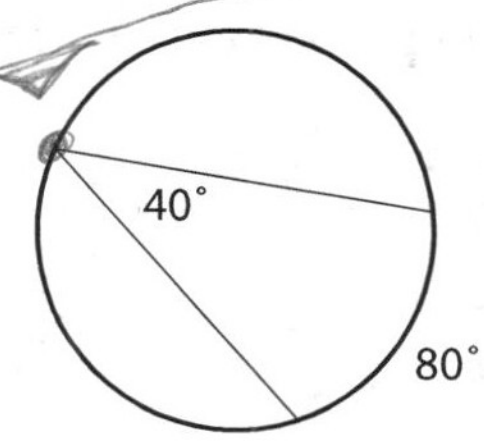

LIFELINE 88 | **When the angle is on the circle, the arc has twice the degree measure.**

A triangle inscribed within a circle, with one leg on the diameter of the circle, creates a 180° arc.

Because the angle measure is ½ the arc measure, the angle is 90°, making the triangle a right triangle.

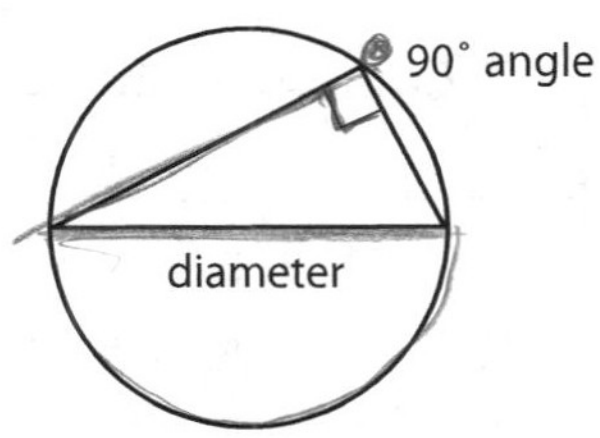

QUIZ | **ANGLES ON A CIRCLE**

The triangle shown is inscribed within the circle. One side of the triangle is on the diameter of the circle. What's the area of the triangle?

(A) 25
(B) 26
(C) 30
(D) 32.5
(E) 40

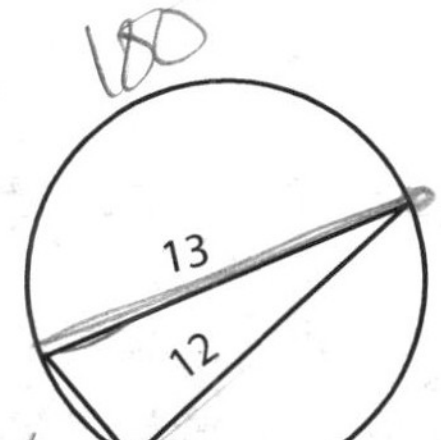

SOLUTION NEXT PAGE

GEOMETRY | QUIZ | ANGLES ON A CIRCLE

CIRCLES

The triangle shown is inscribed within the circle. One side of the triangle is on the diameter of the circle. What's the area of the triangle?

(A) 25
(B) 26
(C) 30
(D) 32.5
(E) 40

SOLUTION

STEP 1 Separate the information from the question

Copy down the drawing

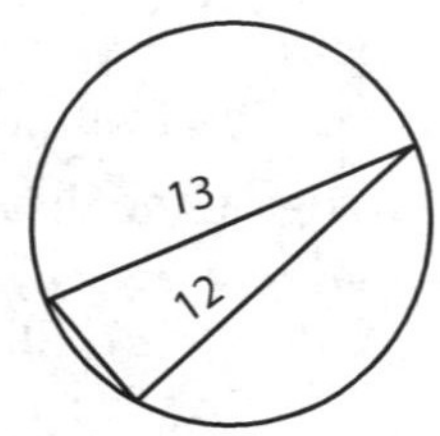

STEP 2 Turn this into something you can use

For more on common right triangles, see p. 223.

One side of the triangle is on the diameter of the circle, making this a right triangle.

This is a 5 : 12 : 13 triangle, making the third side 5.

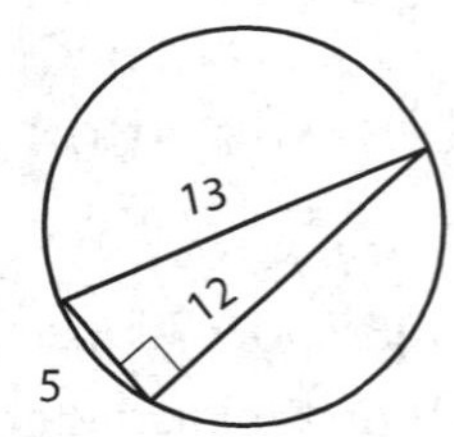

STEP 3 Use this to answer the question

The triangle sides adjacent to the right angle are the base and height.

$$\frac{5 \times 12}{2} = 30$$

Answer is C

TOPIC — SHAPES WITH SAME PERIMETER

GEOMETRY

GEOMETRIC RATIOS

Geometric ratio questions have two shapes with either the same area or the same perimeter.

LIFELINE 89

When the shapes have the same perimeter, set the perimeters equal to each other.

If a circle and a square have the same perimeter.

A circle's perimeter is $2\pi r$
A square's perimeter is $4s$

1. Set the perimeters equal to each other — $2\pi r = 4s$
2. To find the ratio, simplify by dividing both sides by 2 — $\pi r = 2s$

3a. For the ratio based on r: divide both sides by π — $r = \frac{2s}{\pi}$

3b. For the ratio based on s: divide both sides by 2 — $\frac{\pi r}{2} = s$

QUIZ — SHAPES WITH SAME PERIMETER

The perimeters of a square and a rectangle are equal. If the sides of the rectangle are in the ratio of 1 : 2, what is the ratio of the area of the rectangle to the area of the square?

(A) 11 : 10
(B) 8 : 9
(C) 6 : 7
(D) 3 : 4
(E) 1 : 2

SOLUTION NEXT PAGE

GEOMETRY	QUIZ	SHAPES WITH SAME PERIMETER

GEOMETRIC RATIOS

The perimeters of a square and a rectangle are equal. If the sides of the rectangle are in the ratio of 1 : 2, what is the ratio of the area of the rectangle to the area of the square?

(A) 11 : 10
(B) 8 : 9
(C) 6 : 7
(D) 3 : 4
(E) 1 : 2

SOLUTION

STEP 1 | **Separate the information from the question**

The perimeters of a square and a rectangle are equal, and the sides of the rectangle are in the ratio of 1 : 2

STEP 2 | **Turn this into something you can use**

Draw the pictures and label the sides of the rectangle with the given ratio.

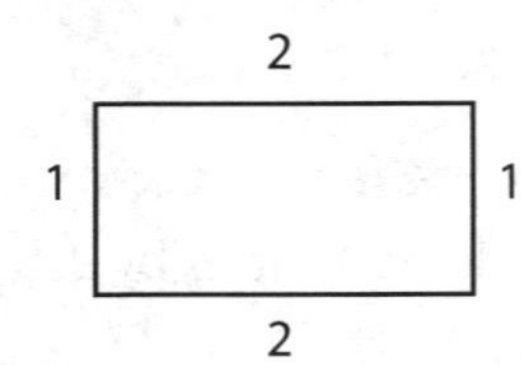

The rectangle thus has a perimeter of 6.

Because the perimeters of the two shapes are equal, the square perimeter is also 6. Label the sides.

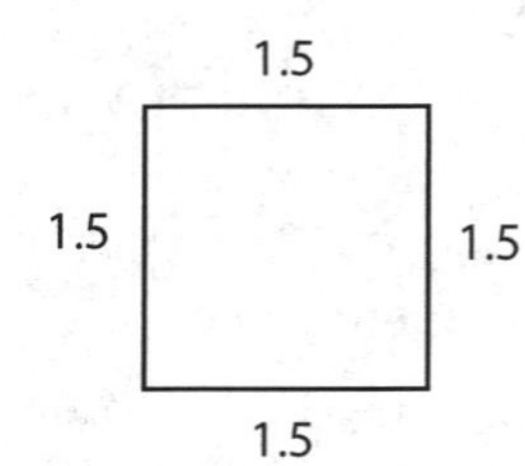

STEP 3 | **Use this to answer the question**

The area of the rectangle is 2, and the area of the square is 2.25, for a ratio of 2 : 2.25.

Ratios use whole numbers, so multiply both sides by 4 for a ratio of 8 : 9

Answer is B

LIFELINE 90

When the shapes have the same area, set the areas equal to each other.

GEOMETRIC RATIOS

If a circle and a square have the same area

A circle's area is πr^2

A square's area is s^2

1. Set the areas equal to each other — $s^2 = \pi r^2$
2. For the ratio based on s: take the square root of both sides. — $s = r\sqrt{\pi}$
3. For the ratio based on r: divide both sides by $\sqrt{\pi}$ — $\frac{s}{\sqrt{\pi}} = r$

QUIZ

SHAPES WITH SAME AREA

If a circle with a radius r has the same area as a square with side s, what is r in terms of s?

(A) $r = \frac{s}{\sqrt{\pi}}$

(B) $r = \frac{s}{\pi}$

(C) $r = \frac{s^2}{\pi}$

(D) $r = \frac{s^2}{\sqrt{\pi}}$

(E) $r = s\pi$

SOLUTION NEXT PAGE

GEOMETRY

QUIZ — SHAPES WITH SAME AREA

GEOMETRIC RATIOS

If a circle with a radius r has the same area as a square with side s, what is r in terms of s ?

(A) $r = \frac{s}{\sqrt{\pi}}$

(B) $r = \frac{s}{\pi}$

(C) $r = \frac{s^2}{\pi}$

(D) $r = \frac{s^2}{\sqrt{\pi}}$

(E) $r = s\pi$

SOLUTION

STEP 1 Separate the information from the question

A circle with radius r has the same area as a square with side s

STEP 2 Turn this into something you can use

Draw the pictures, labeling the radius r and side s.

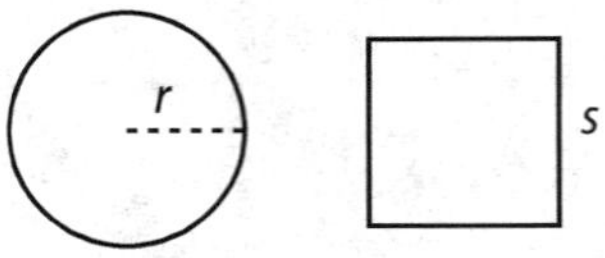

STEP 3 Use this to answer the question

Write an equation setting the two areas equal to each other.

The areas are the same, so set up this equation: $\pi r^2 = s^2$

Isolate the r:

1. Divide both sides by π $r^2 = \frac{s^2}{\pi}$

2. Square root both sides $\sqrt{r^2} = \sqrt{\frac{s^2}{\pi}}$

3. And simplify $r = \frac{s}{\sqrt{\pi}}$

Answer is A

Remember
The area of the circle can be found with $A = \pi r^2$

The area of the square can be found with $A = s^2$

thinktankSCHOLASTIC

TOPIC RECTANGULAR SOLIDS

GEOMETRY

3-D SHAPES

The **GMAT rectangular solid** is a perfect shoebox: each side is a rectangle, and opposite sides are parallel.

LIFELINE 91

When working with the rectangular solid, remember that all the sides are rectangles.

Step	**Example**
Multiply the length, width, and height	$2 \times 3 \times 5$ Volume is 30

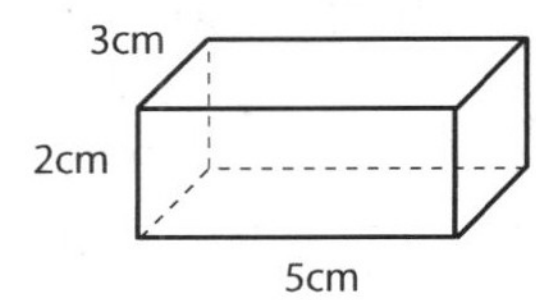

Finding the surface area:

Measure each rectangle	
Top and bottom (same)	$2(3 \times 5) = 30$
Front and back (same)	$2(2 \times 5) = 20$
Left and right sides (same)	$2(2 \times 3) = 12 +$
	Surface area is 62

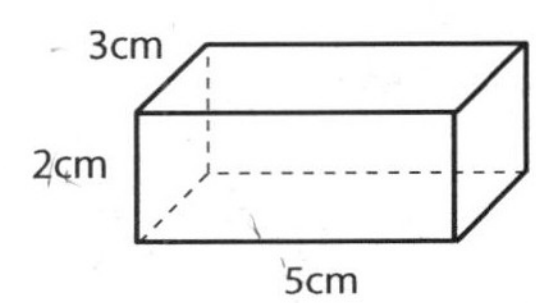

DRILL RECTANGULAR SOLIDS

What are the volume and the surface area of this rectangular solid?

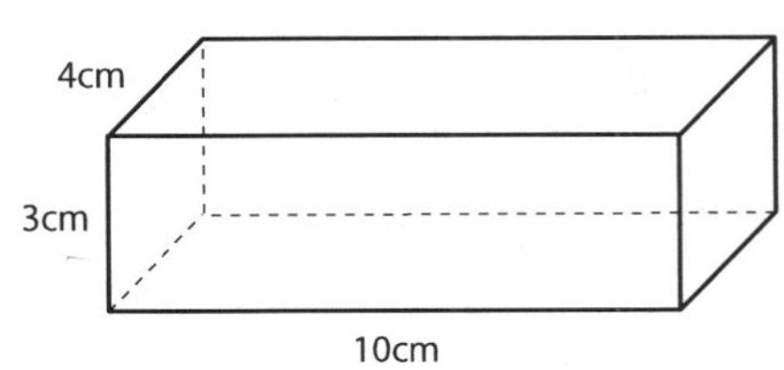

4x3x10 = 120

3x10x2　　2(3·10) +
4x10x2　　2(4·10) +
4x3x2　　2(4·3) +

SA

SOLUTION NEXT PAGE

GEOMETRY | DRILL | RECTANGULAR SOLIDS

3-D SHAPES

What are the volume and surface area of this rectangular solid?

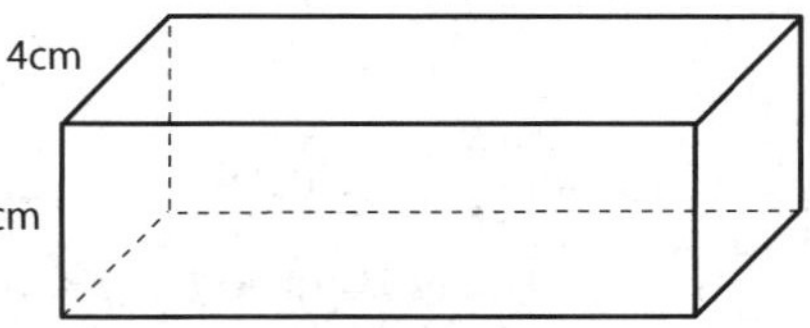

Finding the volume:

SOLUTION

Step	**Example**
Multiply the length, width, and height	$3 \times 4 \times 10$. Volume is 120

Finding the surface area:

Measure each rectangle

Top and bottom (are the same)	$2(4 \times 10) = 2(40) = 80$
Front and back (also the same)	$2(3 \times 10) = 2(30) = 60$
Left and right sides (still the same)	$2(4 \times 3) = 2(12) = 24 +$
Total the measurements:	Surface area is 164

If the rectangular box shown has a volume of 60 cubic inches, what is the area of rectangle *ABCD* ?

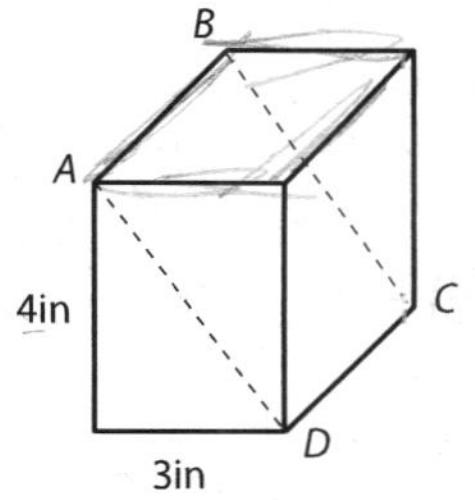

(A) 12
(B) 24
(C) 25
(D) 30
(E) 32

SOLUTION NEXT PAGE

GEOMETRY | QUIZ | RECTANGULAR SOLIDS

3-D SHAPES

If the rectangular box shown has a volume of 60 cubic inches, what is the area of rectangle *ABCD* ?

SOLUTION

STEP 1 | **Separate the information from the question**

The rectangular box has a volume of 60 cubic inches.

STEP 2 | **Turn this into something you can use**

Copy down the drawing

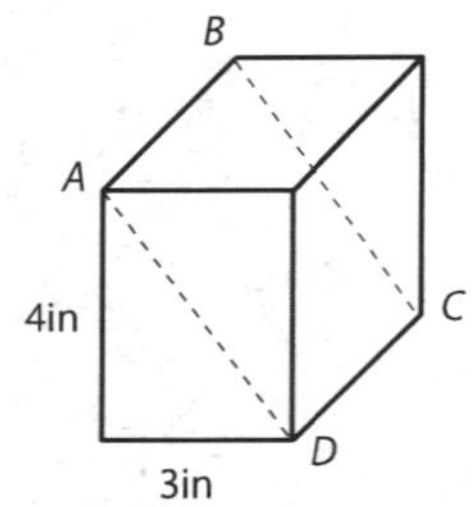

STEP 3 | **Use this to answer the question**

To find the area of rectangle *ABCD*, you need the rectangle's base (*CD*) and height (*AD*).

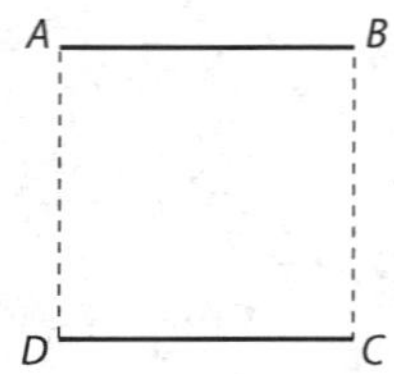

First find the base *CD*: Because volume = length × width × height, the missing edge *AB* (and *CD*) can be found with this equation:

$$\begin{aligned} \text{length} \times 3 \times 4 &= 60 \\ \text{length} \times 12 &= 60 \\ \text{length} &= 5 \end{aligned}$$

For more on common right triangles, see p. 223.

Now find the height *AD*: Because the box is made of rectangles, the dotted segment *AD* creates a right triangle. This is a 3-4-5 triangle, making dotted segment *AD* = 5.

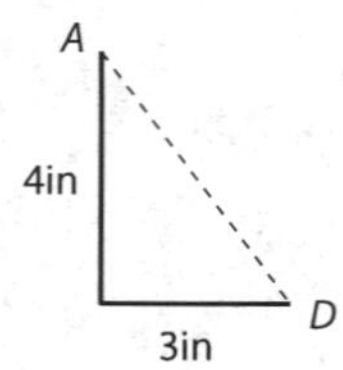

The base and height of the rectangle *ABCD* are 5 and 5, respectively, making the area 25.

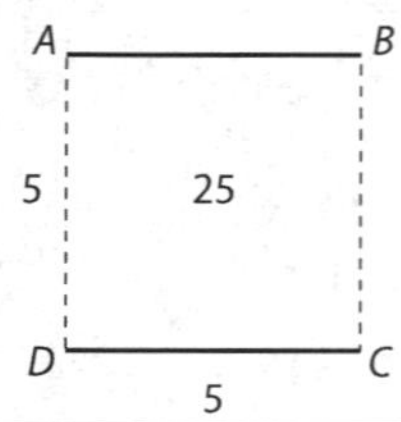

Answer is C

TOPIC **THE CUBE** | **GEOMETRY**

3-D SHAPES

The **cube** is a specific rectangular solid where each side, or face, is a square. Because the **length**, **width**, and **height** are the same, each is called an **edge**.

LIFELINE 92

When working with the cube, use the *edge* formulas.

Finding the volume of a cube:

Step	**Example**	**Formula**
Cube the edge	$3^3 = 27$	$edge^3$

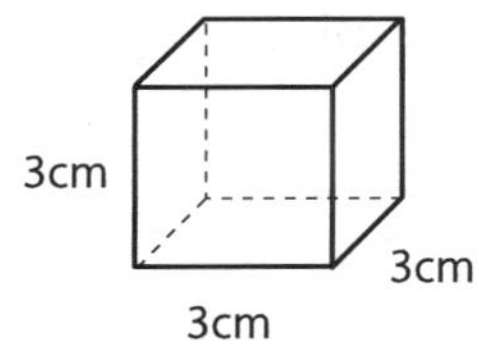

Finding the surface area of a cube:

Steps	**Example**	**Formula**
1. Square the edge (for the area of the face)	$3^2 = 9$	
2. Multiply by 6 (because there are six faces)	$6 \times 9 = 54$	$6(edge^2)$

DRILL **THE CUBE**

Find the volume and surface area of a cube with an edge length of 2.

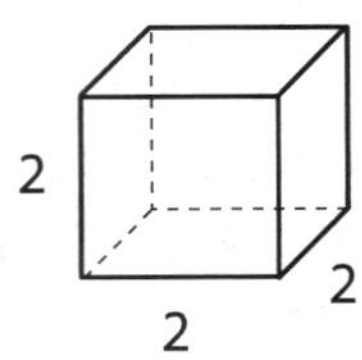

QUIZ **THE CUBE**

If the volume of a cube is 1, what is the surface area of the cube?

(A) 1
(B) 2
(C) 3
(D) 5
(E) 6

SOLUTION NEXT PAGE

GEOMETRY

3-D SHAPES

DRILL THE CUBE

SOLUTION

Find the volume and surface area of a cube with an edge length of 2.

Volume = $edge^3$

$= 2^3$

$= 8$

Surface area = $6(edge^2)$

$= 6(2^2)$

$= 6(4)$

$= 24$

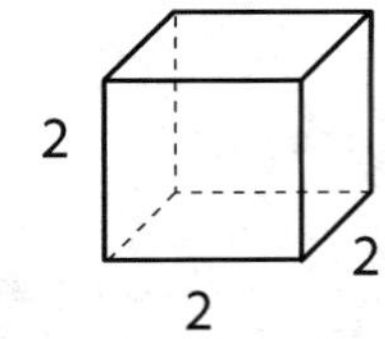

QUIZ THE CUBE

If the volume of a cube is 1, what is the surface area of the cube?

SOLUTION

STEP 1 Separate the information from the question

The volume of a cube is 1

STEP 2 Turn this into something you can use

Draw a cube with a volume of 1

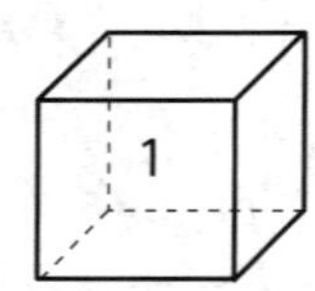

STEP 3 Use this to answer the question

To find the surface area, use the edge length.
Find this by backsolving from the volume.

Volume = $edge^3$	$1 = edge^3$
Cube root both sides	$1 = edge$

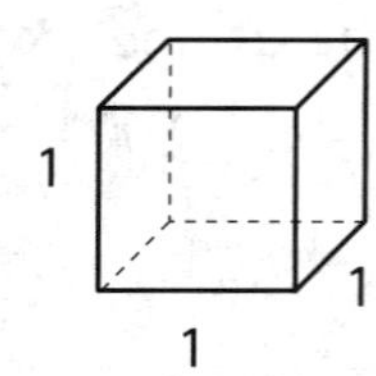

And each edge has a length of 1.

Knowing this length, find the surface area.

Surface area = $6(edge^2)$	Surface area = $6(1^2)$
And simplify	Surface area = 6

Answer is E

TOPIC **THE CYLINDER**

GEOMETRY

3-D SHAPES

The GMAT cylinder is a perfect can of soup: the top and bottom are circular, they are parallel to each other, and the curved surface (i.e., the can label) is perpendicular to the circles.

This is called a **right circular cylinder**.

Calculating the volume is more common than calculating the surface area.

LIFELINE 93

When working with the cylinder, the volume is the circle's area times the cylinder's height.

Draw a picture to organize your thoughts.

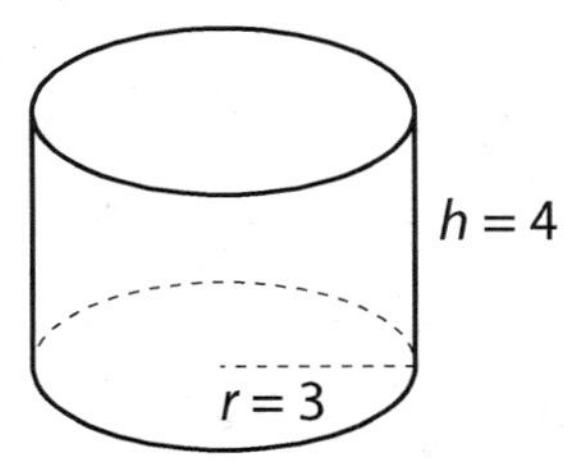

Finding the volume

Steps	**Example**	**Formula**
Multiply the area of one of the circles by the height.	$\pi(3^2)4$ Volume is 36π	$\pi r^2 h$

Finding the surface area

Measure the circles

Top Circle $\pi r^2 = 9\pi$

Bottom Circle $\pi r^2 = 9\pi$

Measure the curved surface (the label of the can), which is a rectangle

Height is 4 Base is the circumference of the circle

Area is 24π

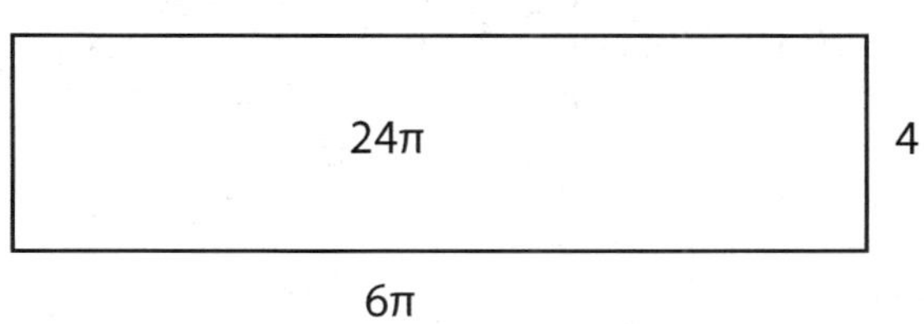

Total the measurements

$$\begin{array}{r} 9\pi \\ 9\pi \\ +\ 24\pi \\ \hline 42\pi \end{array}$$

Surface area is 42π

GEOMETRY

3-D SHAPES

DRILL

THE CYLINDER

Find the volume and surface area of a right circular cylinder with a radius of 4 and a height of 5.

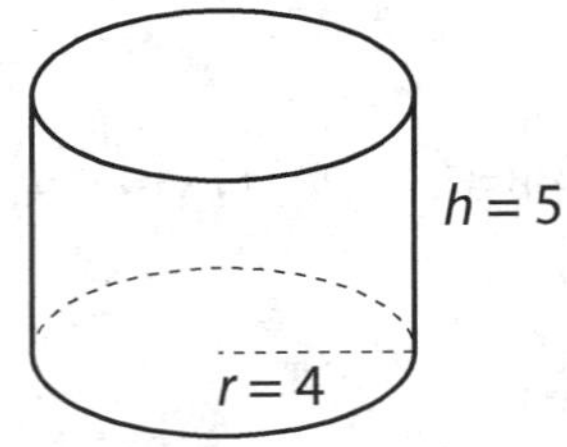

SOLUTION

Finding the volume	**Example**	**Formula**
Multiply the area of one of the circles by the height.	$\pi(4^2)5$ Volume is 80π	$\pi r^2 h$

Finding the surface area

Measure the circles

Top Circle $\pi r^2 = 16\pi$

Bottom Circle $\pi r^2 = 16\pi$

Measure the curved surface (the label of the can), which is a rectangle:

Height is 5
Base is the circumference of the circle

Area is 40π

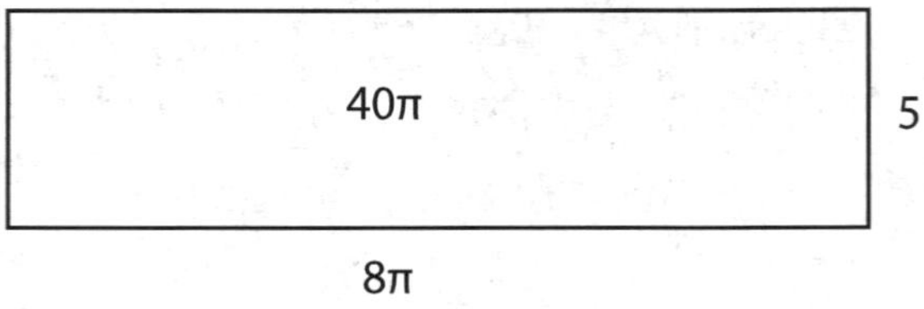

Total the measurements

$$\begin{array}{r} 16\pi \\ 16\pi \\ +\ 40\pi \\ \hline 72\pi \end{array}$$

Surface area is 72π

 thinktankSCHOLASTIC

If a rectangular block that is 1 inch by 2 inches by 5 inches is placed inside a right circular cylinder of radius 2 inches and height 6 inches, how many cubic inches of volume are in the unoccupied portion of the cylinder?

(A) 18π
(B) 24π
(C) $18\pi - 10$
(D) $21\pi - 10$
(E) $24\pi - 10$

SOLUTION NEXT PAGE

GEOMETRY | QUIZ | THE CYLINDER

3-D SHAPES

If a rectangular block that is 1 inch by 2 inches by 5 inches is placed inside a right circular cylinder of radius 2 inches and height 6 inches, how many cubic inches of volume are in the unoccupied portion of the cylinder?

SOLUTION

STEP 1 Separate the information from the question

A rectangular block that is 1 inch by 2 inches by 5 inches is placed inside a cylinder of radius 2 inches and height 6 inches.

STEP 2 Turn this into something you can use

It's basically, "A whiteboard eraser is placed inside of a tennis ball can," but draw the shapes separately.

STEP 3 Use this to answer the question

The difference between two shapes

Remember
To find the surrounding area of two shapes, subtract the smaller shape from the larger shape.

Steps	**Example**	
1. Draw the two shapes separately	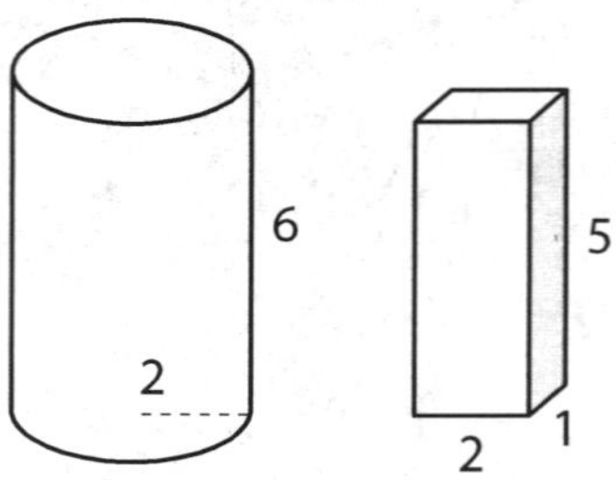	
2. Find the volume of each shape	**Cylinder**	**Rectangular**
	$\pi r^2 h$	$L \times w \times h$
	24π	10
3. Subtract the smaller shape from the larger shape	$24\pi - 10$	

Answer is E

TOPIC | **OTHER 3-D SHAPES**

You may encounter other shapes, such as the sphere, the cone, and the prism. You won't have to measure the volume or the surface area.

LIFELINE 94 | **When working with a different 3-D shape, draw a picture to see how it fits with the basic shapes.**

You don't need to know the formulas for these other shapes, but you do need to recognize how they fit with the other shapes covered in this lesson.

QUIZ | **OTHER 3-D SHAPES**

What's the radius of the largest sphere that can be placed inside a cube of volume 64?

(A) 16
(B) 8
(C) 4
(D) 3
(E) 2

SOLUTION NEXT PAGE

GEOMETRY	QUIZ	OTHER 3-D SHAPES

3-D SHAPES

What's the radius of the largest sphere that can be placed inside a cube of volume 64?

(A) 16
(B) 8
(C) 4
(D) 3
(E) 2

SOLUTION

STEP 1 **Separate the information from the question**

The cube has a volume of 64

STEP 2 **Turn this into something you can use**

If the cube's volume is 64, then each side is 4 wide.

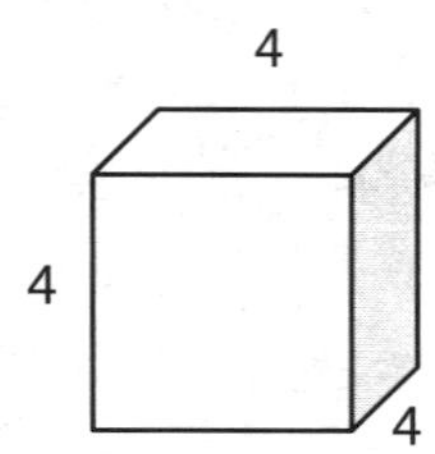

STEP 3 **Use this to answer the question**

The largest circle or sphere that fits inside this cube has a diameter of 4.

The radius is half the diameter, so the radius is 2.

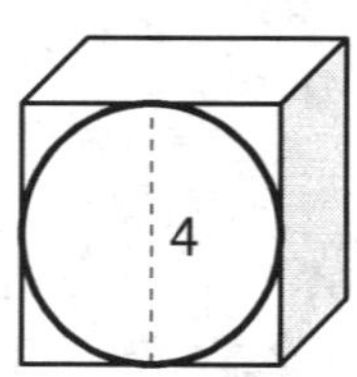

Answer is E

HOMEWORK	GEOMETRY
13th Editon *GMAT Review Book,* pp. 152 -185, question numbers	GMAT REVIEW

ANSWERS NEXT PAGE

Triangles
75 __
92 __
104 __
161 __
165 __
210 __

Quadrilaterals
3 __
13 __
62 __
121 __
147 __
159 __
166 __

Circles
36 __
69 __
175 __
213 __

Geometric Ratios
120 __

3-Shapes
18 __
78 __

GEOMETRY	HOMEWORK		
GMAT REVIEW	**13th Editon** *GMAT Review Book,* pp. 152 -185, question numbers		
ANSWERS	**Triangles** 75 __ D 92 __ D 104 __ B 161 __ A 165 __ E 210 __ D **Quadrilaterals** 3 __ E 13 __ A 62 __ A 121 __ D 147 __ B 159 __ C 166 __ A **Circles** 36 __ C 69 __ E 175 __ D 213 __ B **Geometric Ratios** 120 __ D **3-Shapes** 18 __ D 78 __ A		

Word Problems

Word Problems

Lesson 4

A word problem is any math question embedded within a story. For example, "if 5 horses are in a corral, and 3 run out, how many are left?" is a word problem.

We work with this type of question throughout the book. However, this lesson covers word problems based on specific formulas.

Some of the question types here are among the most commonly-occurring on the GMAT.

TOPIC | **INDIVIDUALS—DIFFERENT RATES** | **WORD PROBLEMS**

WORK PROCESS

The GMAT asks for the time required for a group of people or machines (hereafter "individuals") to complete a shared task.

One type of work problem has individuals working at **different** rates.

LIFELINE 95

With individuals working at different rates, find the fraction of the job each individual does in one hour or day.

Example: If it takes Pat 5 hours to mow a lawn, and it takes Chris 3 hours to mow the same lawn, then working together, how long will it take them to mow this lawn?

Steps	Example
1. Find the fraction of the job that each individual works in one hour.	Pat mows $\frac{1}{5}$ of the lawn in 1 hour. Chris mows $\frac{1}{3}$ of the lawn in 1 hour.
2. Add the fractions to find the portion of the job that the two of them, together, work in one hour.	$\left[\frac{1}{5}+\frac{1}{3}\right]=\left[\frac{3}{15}+\frac{5}{15}\right]=\frac{8}{15}$ Amount of the job done in 1 hour.
3. Take the **reciprocal** of that fraction to determine the time required to work the entire job.	$\frac{8}{15} \rightarrow \frac{15}{8} = 1\frac{7}{8}$ hours Amount of time required to do work the entire job.

5 hours to mow the lawn means 1/5 gets mowed in 1 hour.

You will never have to convert your answer to "1 hour, 52 minutes, 30 seconds."

DRILL | **INDIVIDUALS—DIFFERENT RATES**

If Sam can paint a certain house in 2 days, Alex can paint the same house in 4 days, and Robin can paint the house in 5 days, then working together, how many days are needed for them to paint the house?

WORD PROBLEMS | **DRILL** | **INDIVIDUALS—DIFFERENT RATES**

WORK PROCESS

If Sam can paint a certain house in 2 days, Alex can paint the same house in 4 days, and Robin can paint the house in 5 days, then working together, how many days are needed for them to paint the house?

SOLUTION

Steps	**Example**
1. Find the fraction of the job that each individual does in one day.	Sam, Alex, and Robin each paints $\frac{1}{2}, \frac{1}{4}$, and $\frac{1}{5}$ of the house
2. Add the fractions to find the fraction of the house that is painted in one day when they work together.	$\frac{1}{2} + \frac{1}{4} + \frac{1}{5} =$ $\frac{10}{20} + \frac{5}{20} + \frac{4}{20} = \frac{19}{20}$
3. Take the **reciprocal** of that fraction to determine the time required to paint the whole house.	$\frac{19}{20} \rightarrow \frac{20}{19} = 1\frac{1}{19}$ days

DETAIL | **RECIPROCATING THE FINAL FRACTION—EXAMPLES**

The idea of reciprocating the fraction to find the time required can be tricky to wrap your mind around. Think of a timeline.

If $\frac{1}{3}$ of a job is done in an hour, it will take 3 hours to complete the job.

Steps	**Example**
1. Start with the fraction of the job of the job that is done in an hour	$\frac{1}{3}$
2. Take the reciprocal of that fraction	$\frac{3}{1}$
3. And you have the time needed	3 hours

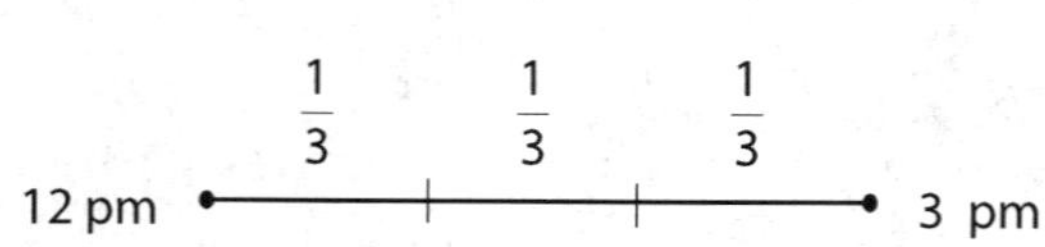

DETAIL	RECIPROCATING THE FINAL FRACTION—EXAMPLES

WORK PROCESS

Here's another example: Suppose I work $\frac{2}{5}$ of a job in one hour.

From 12pm to 1pm, I work $\frac{2}{5}$ of the job.

From 1pm to 2pm, another $\frac{2}{5}$

From 2pm to 2:30pm, the remaining $\frac{1}{5}$

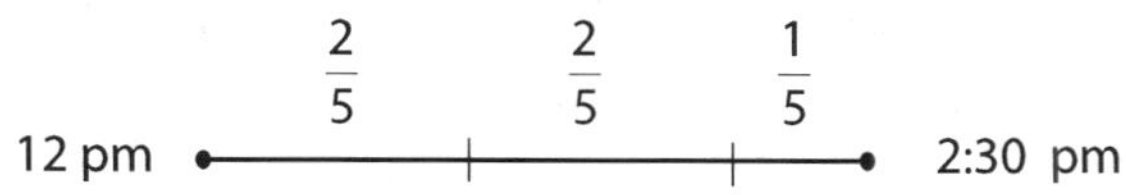

It took me $2\frac{1}{2}$, or $\frac{5}{2}$ hours to complete the job.

$\frac{5}{2}$ is the reciprocal of $\frac{2}{5}$

QUIZ	INDIVIDUALS—DIFFERENT RATES

The water from one outlet, flowing at a constant rate, can fill a city water tank in 10 hours. The tank drain, flowing at a constant rate, can empty the tank in 15 hours. If the tank drain is left open while the water outlet fills the tank, approximately what is the number of hours required to fill the tank?

(A) 30
(B) 20
(C) 15
(D) 10
(E) 5

SOLUTION NEXT PAGE

WORD PROBLEMS | **QUIZ** | **INDIVIDUALS—DIFFERENT RATES**

WORK PROCESS

The water from one outlet, flowing at a constant rate, can fill a city water tank in 10 hours. The tank drain, flowing at a constant rate, can empty the tank in 15 hours. If the tank drain is left open while the water outlet fills the tank, approximately what is the number of hours required to fill the tank?

(A) 30
(B) 20
(C) 15
(D) 10
(E) 5

SOLUTION

STEP 1 | **Separate the information from the question**

The water from one outlet can fill a city water tank in 10 hours.

The tank drain can empty the tank in 15 hours.

STEP 2 | **Turn this into something you can use**

$\frac{1}{10}$ of the tank gets filled in one hour, while

$\frac{1}{15}$ of the tank is emptied in the same hour.

STEP 3 | **Use this to answer the question**

In each hour, while $\frac{1}{10}$ gets filled, $\frac{1}{15}$ gets emptied.

Steps	Example
1. Start with the fraction of the job that each individual does in one hour.	$\frac{1}{10}$ and $-\frac{1}{15}$
2. Add (or in this case, subtract) the fractions.	$\frac{3}{30} - \frac{2}{30} = \frac{1}{30}$
3. Take the reciprocal of the result	$\frac{1}{30} \rightarrow \frac{30}{1}$
The time needed is	30 hours

Answer is A

The other type of work problem has individuals working at the **same** rate.

LIFELINE 96

With individuals working at the same rate, find the fraction of the job one individual does in one hour or day.

Example: If it takes 6 people 12 hours to do a job

1) How many people are needed to do the job in 8 hours?

Steps	**Example**
1. Find the fraction of the job that one individual does in one hour.	$\frac{1}{\text{\#hours / days}} \times \frac{1}{\text{\#individuals}}$
If it takes the team 12 hours, then 1/12 is done in one hour.	$\frac{1}{12}$
On a 6-person team, each person does 1/6 of the work.	$\frac{1}{6}$ $\frac{1}{12} \times \frac{1}{6} = \frac{1}{72}$
One person does 1/72 of the job in one hour.	$\frac{1}{72}$
2. Multiply the fraction by the new number of hours.	$\frac{1}{72} \times \frac{8}{1} = \frac{8}{72} = \frac{1}{9}$
3. Take the reciprocal.	$\frac{1}{9} \rightarrow \frac{9}{1}$

Answer: 9 people are needed to get the job done in 8 hours.

2) How many hours will it take 4 people?

1. Find the fraction of the job that one individual does in one hour.	$\frac{1}{12} \times \frac{1}{6} = \frac{1}{72}$
2. Multiply the fraction by the new number of people.	$\frac{1}{72} \times \frac{4}{1} = \frac{4}{72} = \frac{1}{18}$
3. Take the reciprocal.	$\frac{1}{18} \rightarrow \frac{18}{1}$

Answer: 4 people can get the job done in 18 hours.

WORD PROBLEMS | **DRILL 1** | **INDIVIDUALS—SAME RATE**

WORK PROCESS

If it takes 10 machines 6 days to do a certain job

1) How many machines are needed to do the job in 2 days?

2) How many days will it take 12 machines to do the job?

SOLUTION

1) How many machines are needed to do the job in 2 days?

Steps	**Example**
1. Find the fraction of the job that one individual does in one day	$\frac{1}{\text{\#hours / days}} \times \frac{1}{\text{\#individuals}}$
If it takes the team 6 days, then 1/6 is done in one day.	$\frac{1}{6}$
On a 10-machine team, each machine does 1/10 of the work.	$\frac{1}{10}$ $\frac{1}{6} \times \frac{1}{10} = \frac{1}{60}$
One machine does 1/60 of the job in one day.	$\frac{1}{60}$
2. Multiply this by the new number of days	$\frac{1}{60} \times \frac{2}{1} = \frac{2}{60} = \frac{1}{30}$
3. Take the reciprocal	$\frac{1}{30} \rightarrow \frac{30}{1}$

30 machines are needed to do the job in 2 days.

2) How many days will it take 12 machines to do the job?

1. Find the fraction of the job that one individual does in one day	$\frac{1}{6} \times \frac{1}{10} = \frac{1}{60}$
2. Multiply by the new number of individuals	$\frac{1}{60} \times \frac{12}{1} = \frac{12}{60} = \frac{1}{5}$
3. Take the reciprocal	$\frac{1}{5} \rightarrow \frac{5}{1}$

5 days are needed for 12 machines to do the job.

DRILL 2	INDIVIDUALS—SAME RATE	WORD PROBLEMS

WORK PROCESS

If it takes 50 people 10 hours to do a certain job:

1) How many hours will it take 20 people?

2) How many people are needed to do the job in 4 hours?

SOLUTION NEXT PAGE

WORD PROBLEMS	DRILL 2	INDIVIDUALS—SAME RATE

WORK PROCESS

If it takes 50 people 10 hours to do a certain job:

SOLUTION

1) How many hours will it take 20 people?

Steps	Example
1. Find the fraction of the job that one individual does in one day	$\frac{1}{\#\text{hours / days}} \times \frac{1}{\#\text{individuals}}$
If it takes the team 10 hours, then 1/10 is done in one hour.	$\frac{1}{10}$
On a 50-person team, each machine does 1/50 of the work.	$\frac{1}{50}$ $\frac{1}{10} \times \frac{1}{50} = \frac{1}{500}$
One machine does 1/500 of the job in one hour.	$\frac{1}{500}$
2. Multiply this by the new number of individuals.	$\frac{1}{500} \times \frac{20}{1} = \frac{20}{500} = \frac{1}{25}$
3. Take the reciprocal	$\frac{1}{25} \rightarrow \frac{25}{1}$

Answer: It takes 25 hours for 20 people to complete the job.

2) How many people are needed to do the job in 4 hours?

Steps	Example
1. Find the fraction of the job that one individual does in one hour	$\frac{1}{10} \times \frac{1}{50} = \frac{1}{500}$
2. Multiply by the new number of individuals	$\frac{1}{500} \times \frac{4}{1} = \frac{4}{500} = \frac{1}{125}$
3. Take the reciprocal	$\frac{1}{125} \rightarrow \frac{125}{1}$

Answer: 125 individuals are needed to complete this job in 4 hours.

If it takes 15 machines 21 days to do a certain job:

1) How many days will it take 9 machines?

2) How many machines are needed to do the job in 3 days?

SOLUTION NEXT PAGE

WORD PROBLEMS | **DRILL 3** | **INDIVIDUALS—SAME RATE**

WORK PROCESS

If it takes 15 machines 21 days to do a certain job:

SOLUTION

1) How many days will it take 9 machines?

Steps	Example
1. Find the fraction of the job that one individual does in one day	$\frac{1}{\text{\#hours / days}} \times \frac{1}{\text{\#individuals}}$
If it takes the team 21 days, then 1/21 is done in one day.	$\frac{1}{21}$
On a 15-machine team, each machine does 1/15 of the work.	$\frac{1}{15}$ $\frac{1}{21} \times \frac{1}{15} = \frac{1}{315}$
One machine does 1/315 of the job in one hour.	$\frac{1}{315}$
2. Multiply this by the new number of individuals.	$\frac{1}{315} \times \frac{9}{1} = \frac{9}{315} = \frac{1}{35}$
3. Take the reciprocal	$\frac{1}{35} \rightarrow \frac{35}{1}$

Answer: It takes 35 days for 9 machines to complete the job.

2) How many machines are needed to do the job in 3 days?

1. Find the fraction of the job that one individual does in one day.	$\frac{1}{21} \times \frac{1}{15} = \frac{1}{315}$
2. Multiply by the required number of days.	$\frac{1}{315} \times \frac{3}{1} = \frac{3}{315} = \frac{1}{105}$
3. Take the reciprocal.	$\frac{1}{105} \rightarrow \frac{105}{1}$

105 machines are needed to get the job done in 3 days.

QUIZ | WORK PROCESS

WORD PROBLEMS

WORK PROCESS

A company uses two types of machines, type A and type B. Machine A does a certain job in 6 hours, and machine B does the same job in 12 hours. If the company uses the same number of both types of machines, how many total machines are needed to do the job in 2 hours?

(A) 2
(B) 4
(C) 6
(D) 8
(E) 10

SOLUTION NEXT PAGE

WORD PROBLEMS | **QUIZ** | **WORK PROCESS**

WORK PROCESS

A company uses two types of machines, type A and type B. Machine A does a certain job in 6 hours, and machine B does the same job in 12 hours. If the company uses the same number of both types of machines, how many total machines are needed to do the job in 2 hours?

SOLUTION

STEP 1 Separate the information from the question

Machine A does a certain job in 6 hours, and machine B does the same job in 12 hours.

STEP 2 Turn this into something you can use

In one hour, machine A does $\frac{1}{6}$ of the job, and machine B does $\frac{1}{12}$ of the job.

STEP 3 Use this to answer the question

Because the company is using the same number of Machines A and B, consider a **team** of these machines to be the "individual" in the formula.

Steps	Example
1. Find the fraction of the job that one "team A and B" does in one hour	$\frac{1}{6} + \frac{1}{12} = \frac{3}{12} = \frac{1}{4}$ Together they do $\frac{1}{4}$ of the job.
2. Multiply by the new number of "teams A and B"	$\frac{1}{4} \times \frac{2}{1} = \frac{2}{4} = \frac{1}{2}$
3. Take the reciprocal	$\frac{1}{2} \rightarrow \frac{2}{1}$

4. 2 teams are needed to do the job in 2 hours.

5. Each team has two machines, so with 2 teams, there are 4 machines total.

Answer is B

TOPIC	RATE / DISTANCE

On the GMAT, the rate of travel should be written as a fraction, not in terms of mph:

$\frac{50 \textit{ miles}}{1 \textit{ hour}}$ not "50 mph"

LIFELINE 97	**To find the average speed, place the entire distance over the entire time (in hours).**

Rate Formula

$$\text{Average Speed} = \frac{\text{Total Distance}}{\text{Total Time}}$$

Example

$$\frac{200 \textit{ miles}}{4 \textit{ hours}} = \frac{50 \textit{ miles}}{1 \textit{ hour}}$$

is the equivalent of "50 mph"

QUIZ	RATE / DISTANCE

David drove at an average speed of 50 miles per hour for the first 50 miles of a trip and then at an average speed of 20 miles per hour for the remaining 40 miles of the trip. If he made no stops during the trip, what was David's average speed for the entire trip?

(A) 25 mph
(B) 30 mph
(C) 35 mph
(D) 40 mph
(E) 45 mph

SOLUTION NEXT PAGE

WORD PROBLEMS | QUIZ | RATE / DISTANCE

RATE / DISTANCE

David drove at an average speed of 50 miles per hour for the first 50 miles of a trip and then at an average speed of 20 miles per hour for the remaining 40 miles of the trip. If he made no stops during the trip, what was David's average speed for the entire trip?

(A) 25 mph
(B) 30 mph
(C) 35 mph
(D) 40 mph
(E) 45 mph

SOLUTION

STEP 1 Separate the information from the question

David drove at an average speed of 50 miles per hour for the first 50 miles of a trip and then at an average speed of 20 miles per hour for the remaining 40 miles of the trip.

STEP 2 Turn this into something you can use

David drove 60 mph for 50 miles, then 20 mph for 40 miles.

STEP 3 Use this to answer the question

	First leg		**Second leg**		**Total**
Distance:	50 miles	+	40 miles	=	90 miles
Time:	1 hour	+	2 hours	=	3 hours
	(50 mph means he traveled 50 miles in one hour)		(20 mph means he traveled 20 miles in one hour, for 2 hours)		

$$\frac{90\ \textit{miles}}{3\ \textit{hours}} = \frac{30\ \textit{miles}}{1\ \textit{hours}}$$

Answer is B

TOPIC	CONVERSIONS

In a **conversion** question, the GMAT gives you the units of conversion (such as gallons to liters), except for units of time (such as minutes to seconds).

LIFELINE 98	**When working a conversion question, set up the fractions so that you can cancel everything out.**

Example: What is 10 grams per liter in terms of kilograms per milliliter?

(1 kilogram = 1000 grams; 1 liter = 1000 milliliters)

Steps — **Example**

1. Set up the fractions

   a. Start with what the question gives you

$$\left[\frac{10\ grams}{1\ liter}\right]$$

   b. Add each conversion as one fraction, making sure the units are opposite.

$$\left[\frac{10\ grams}{1\ liter}\right]\left[\frac{1\ liter}{1000\ mL}\right]\left[\frac{1\ kg}{1000\ grams}\right]$$

> Never use "m" to abbreviate anything: it could stand for "meters", "minutes", or "miles."
>
> Make sure your lowercase "l" doesn't look like the number "1."

2. Cancel everything you can

   a. Cancel the labels

$$\left[\frac{10\ \not{grams}}{1\ \not{liter}}\right]\left[\frac{1\ \not{liter}}{1000\ mL}\right]\left[\frac{1\ Kg}{1000\ \not{grams}}\right]$$

   b. Cancel the numbers

$$1\left[\frac{\not{10}}{1}\right]\left[\frac{1}{1000\ mL}\right]\left[\frac{1\ Kg}{\not{1000}}\right]100$$

3. Multiply the fractions

$$\left[\frac{1}{1}\right]\left[\frac{1}{1000\ mL}\right]\left[\frac{1\ Kg}{100}\right] = \left[\frac{1\ Kg}{100{,}000\ mL}\right]$$

This question asks for "per milliliter", so convert the fraction to show a single milliliter by moving the decimal points of both the top and the bottom five spaces to the left.

> For more on moving decimals, see Arithmetic, p. 47.

$$\left[\frac{1\ Kg}{100{,}000\ mL}\right] \rightarrow \left[\frac{0.00001\ Kg}{1\ mL}\right]$$

WORD PROBLEMS | **DRILL** | **CONVERSIONS**

CONVERSIONS

If Larry rides at a constant rate of 30 kilometers in one hour, how many meters does he ride in one minute? (1 kilometer = 1000 meters)

SOLUTION

Steps | **Example**

1. Set up the fractions

   a. Start with what the question gives you

$$\left[\frac{30\ km}{1\ sec}\right]$$

   b. Add each conversion as one fraction, making sure the units are opposite.

$$\left[\frac{30\ km}{1\ hour}\right]\left[\frac{1000\ mt}{1\ km}\right]\left[\frac{1\ hour}{60\ min}\right]$$

> **Remember**
> Don't use "m" to abbreviate.

2. Cancel everything you can

   a. Cancel the labels

$$\left[\frac{30\ \cancel{km}}{1\ \cancel{hour}}\right]\left[\frac{1000\ mt}{1\ \cancel{km}}\right]\left[\frac{1\ \cancel{hour}}{60\ min}\right]$$

   b. Cancel the numbers

$$\left[\frac{\overset{1}{\cancel{30}}}{1}\right]\left[\frac{1000\ mt}{1}\right]\left[\frac{1}{\underset{2}{\cancel{60}}\ min}\right]$$

3. Multiply the fractions

$$\left[\frac{1}{1}\right]\left[\frac{1000\ mt}{1}\right]\left[\frac{1}{2\ min}\right]=\left[\frac{1000\ mt}{2\ min}\right]$$

$$=\frac{1000\ mt}{2\ min}=\frac{500\ mt}{1\ min}$$

Larry rides 500 meters in one minute.

A car travels at a constant rate of 25 meters per second. How many kilometers does it travel per minute? (1 km = 1,000 meters)

(A) 1
(B) 1.5
(C) 3
(D) 33.3
(E) 416.7

SOLUTION NEXT PAGE

WORD PROBLEMS | **QUIZ** | **CONVERSIONS**

CONVERSIONS

A car travels at a constant rate of 25 meters per second. How many kilometers does it travel per minute? (1 km = 1,000 meters)

(A) 1
(B) 1.5
(C) 3
(D) 33.3
(E) 416.7

SOLUTION

STEP 1 Separate the information from the question

A car travels at a constant rate of 25 meters per second.

STEP 2 Turn this into something you can use

The car travels $\left[\frac{25\ mt}{1\ sec}\right]$

STEP 3 Use this to answer the question

Steps | **Example**

Remember
Don't use "m" to abbreviate.

1. Set up all the fractions $\left[\frac{25\ mt}{1\ sec}\right]\left[\frac{1\ km}{1000\ mt}\right]\left[\frac{60\ sec}{1\ min}\right]$

2. Cancel everything you can: $\left[\frac{\overset{1}{\cancel{25}}\ \cancel{mt}}{1\ \cancel{sec}}\right]\left[\frac{1\ km}{\cancel{1000}\ \cancel{mt}}\right]\left[\frac{\overset{6}{\cancel{60}}\ \cancel{sec}}{1\ min}\right]$

$1000 \to 100 \cdot 10 \to 4$ (100 and 10 cancelled; 4 remains)

3. Multiply the fractions $\left[\frac{1}{1}\right]\left[\frac{1\ km}{4}\right]\left[\frac{6}{1\ min}\right] = \left[\frac{6\ km}{4\ min}\right]$

$$\frac{3\ km}{2\ min} = \frac{1.5\ km}{1\ min}$$

Answer is B

TOPIC SET DISTRIBUTION

WORD PROBLEMS

SETS

When items are divided into two groups, and there is overlap between the groups, it is a **set distribution** question.

LIFELINE 99

When working on set distribution, use the Set formula.

Example: A school offers History and German classes. There's a group of students in History, a group in German, and the overlap is the students who are taking both classes.

Steps	**Example**
A school offers two classes, History and German.	20 students are in History 18 are in German
The **overlap** is the number of students in both groups	13 students are taking both
Use the formula	Total = $(\text{set}_1 + \text{set}_2) - \text{overlap}_{(1\,\&\,2)}$
Find the total number of students by adding the two classes and subtracting the number of students who are in both (the overlap).	total students = $(20 + 18) - 13 = 25$

If you're given any three items, you can find the fourth.

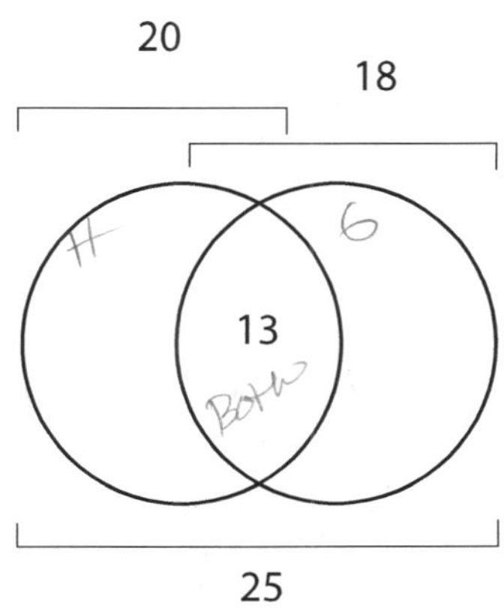

SOLUTION NEXT PAGE

WORD PROBLEMS | **DRILL** | **SET DISTRIBUTION**

SETS

A dealer offers cars with two available options: 150 cars have a sunroofs; 200 cars have satellite radio; 19 cars have both

What is the total number of cars?

SOLUTION

A dealer offers cars with two available options:

150 cars have a sunroofs
200 cars have satellite radio
19 cars have both

What is the total number of cars?

With the formula $\text{Total} = (\text{set}_1 + \text{set}_2) - \text{overlap}_{(1\,\&\,2)}$

Place the numbers into the formula: $\text{Total} = (150 + 200) - 19 = 331$

If b represents the number of tryouts for the baseball team, f represents the number of tryouts for the football team, and $b + f = 120$, how many tryouts are there total if 30 are trying for both teams?

(A 80
(B) 90
(C) 100
(D) 110
(E) 120

SOLUTION NEXT PAGE

WORD PROBLEMS | QUIZ | SET DISTRIBUTION

SETS

If b represents the number of tryouts for the baseball team, f represents the number of tryouts for the football team, and $b + f = 120$, how many tryouts are there total if 30 are trying for both teams?

(A 80
(B) 90
(C) 100
(D) 110
(E) 120

SOLUTION

STEP 1 Separate the information from the question

b represents the number of tryouts for the baseball team
f represents the number of tryouts for the football team
$b + f = 120$, and 30 are trying for both teams

STEP 2 Turn this into something you can use

b = tryouts for baseball
f = tryouts for football
$b + f = 120$
30 = tryouts for both

STEP 3 Use this to answer the question

With the formula	Total = $(\text{set}_1 + \text{set}_2) - \text{overlap}(_{1\,\&\,2})$
Plug in what you have	You don't have sets 1 and 2 separately (baseball and football tryouts), but you do have the sum of the two: 120
This is all you need to answer this question	Total = (120) − 30 = 90

Answer is B

TOPIC SETS IN TABLES

WORD PROBLEMS

SETS

When items are divided into two groups, and there is no overlap between the groups, it is a **set in tables** question.

LIFELINE 100

When working a set in table question, set up the table to do the work.

Example: A certain shelter has only dogs and cats. 40% of the animals are dogs, and half of the animals wear collars. 20% of the total animals are dogs with collars, and 60 dogs have no collars. How many cats wear collars?

Steps

1. Set up the table with the groups on the outside, leaving room to add the numbers.

Example

	Dogs	**Cats**
Collar		
No Collar		

2. Add the percents to the outside.

   If 40% of the animals are dogs, 60% are cats.

	Dogs **40%**	Cats **60%**
Collar **50%**		
No Collar **50%**		

If there are both percents and numbers, work the percents first.

3. Now you need one percent in the inside, and you can solve all of them.

   a. Add the 20% dogs with collars.

	Dogs 40%	Cats 60%
Collar 50%	**20%**	
No Collar 50%		

You also know 60 dogs have no collars, but work all the percents first.

CONTINUED NEXT PAGE

WORD PROBLEMS | LIFELINE | SETS IN TABLES — Continued

SETS IN TABLES

The percents are of the total animals, so everything **in** the table totals 100%.

Steps	Example

b. If 20% of the animals are dogs with collars, 20% are dogs without collars, totalling 40% of dogs. Add that.

	Dogs 40%	Cats 60%
Collar 50%	20%	
No Collar 50%	**20%**	

c. If 20% of the animals are dogs with collars, 30% are cats with collars, totalling 50% of animals with collars. Add that.

	Dogs 40%	Cats 60%
Collar 50%	20%	**30%**
No Collar 50%	20%	

d. The remaining number is 30%, bringing both the cats total to 60% and the no collar total to 50%.

	Dogs 40%	Cats 60%
Collar 50%	20%	30%
No Collar 50%	20%	**30%**

4. Now that all the percents are in the table, use the number given to find the total number of animals.

a. Find the number of animals total:

60 dogs have no collars.
20% of the animals are dogs with no collars
20% = 60 animals

Multiply both sides by 5 to get 100%

100% = 300 animals total.

b. Take the percent from the total:

30% of the animals are cats with collars

30% of 300 is 90.

SETS IN TABLES

An auto dealer's inventory has 30% convertible and 70% hardtop cars. 40% of the cars have manual transmissions, and the rest have automatic. If 50% of the cars are hardtop with automatic transmissions, and 50 cars are hardtop with manual transmissions, how many cars are convertible with manual transmissions?

SOLUTION NEXT PAGE

WORD PROBLEMS | DRILL | SETS IN TABLES

SETS IN TABLES

An auto dealer's inventory has 30% convertible and 70% hardtop cars. 40% of the cars have manual transmissions, and the rest have automatic. If 50% of the cars are hardtop with automatic transmissions, and 50 cars are hardtop with manual transmissions, how many cars are convertible with manual transmissions?

SOLUTION

Steps / **Example**

1. Set up the table with the groups on the outside, leaving room to add the numbers

	Convertible	**Hardtop**
Manual		
Auto		

2. Add the percents to the outside

	Convertible **30%**	Hardtop **70%**
Manual **40%**		
Auto **60%**		

3. Add the percents to the inside (First put the 20% hardtop with manual, then fill in the rest)

	Convertible 30%	Hardtop 70%
Manual 40%	**20%**	**20%**
Auto 60%	**10%**	**50%**

4. Use the number given to find the total number of cars.

a. Find the number of cars total: 20%, which is 50 cars, are hardtop with manual transmissions

Multiply both sides by 5 to get 100%. 100% = 250 cars total

b. Take the percent from the total: 10% of the cars are convertible with manual transmission. 10% of 250 is 25.

Of the 50 candidates for a promotion at the office, 22 have at least 5 years' experience, 14 have MBAs, and 17 have neither the experience nor an MBA. How many of the candidates have at least 5 years' experience and an MBA?

(A) 3
(B) 11
(C) 17
(D) 19
(E) 36

SOLUTION NEXT PAGE

WORD PROBLEMS | QUIZ 1 | SETS IN TABLES

SETS IN TABLES

Of the 50 candidates for a promotion at the office, 22 have at least 5 years' experience, 14 have MBAs, and 17 have neither the experience nor an MBA. How many of the candidates have at least 5 years' experience and an MBA?

SOLUTION

STEP 1 Separate the information from the question

50 candidates total
22 have at least 5 years' experience
14 have MBAs
17 have neither experience nor an MBA

STEP 2 Turn this into something you can use

1. Set up the table with the groups on the outside, leaving room to add the numbers

	Experience	No Experience
MBA		
No MBA		

2. Add the numbers to the outside

	Experience 22	No Experience 28
MBA 14		
No MBA 36		

3. Add the numbers to the inside (first put in the 17 with neither experience nor an MBA.

	Experience 22	No Experience 28
MBA 14		
No MBA 36		17

STEP 3 Use this to answer the question

Everything inside the table totals 50, which is the number of candidates.

Add the rest

	Experience 22	No Experience 28
MBA 14	**3**	11
No MBA 36	19	17

Answer is A

At Malcolm University, 50% of the student body is graduate, and the remaining 50% is undergraduate. 40% of the student body is graduate females. What percent of the graduate students are males?

(A) 10
(B) 20
(C) 30
(D) 40
(E) 50

SOLUTION NEXT PAGE

WORD PROBLEMS | QUIZ 2 | SETS IN TABLES

SETS IN TABLES

At Malcolm University, 50% of the student body is graduate, and the remaining 50% is undergraduate. 40% of the student body is graduate females. What percent of the graduate students are males?

SOLUTION

STEP 1 Separate the information from the question

50% is graduate
50% is undergraduate
40% of the student body is graduate females.

STEP 2 Turn this into something you can use

1. Set up the table with the groups on the outside, leaving room to add the numbers

	Graduate	**Undergrad**
Males		
Females		

2. Add the numbers that you have

	Graduate **50%**	Undergrad **50%**
Males	**10%**	
Females	**40%**	

STEP 3 Use this to answer the question

The question is not, "What percent of the student body is graduating males?" The answer to that is 10%.

If you picked A, 10%, you fell for a trap that many test-takers fall into, and you're in good company.

The question is, "What percent of the graduate students are males?"

10% of 50% is 20%

Answer is B

Devious? Yes. Like the GMAT? Absolutely.

Be sure to stay on your toes and watch for traps, no matter how comfortable you become with the math.

TOPIC	CORE COUNTING METHOD

Counting methods refers to finding the number of possible outcomes.

LIFELINE 101

When finding the number of possible outcomes, start by placing a dash for each event.

served in that order

3 things

Example: A restaurant offers six appetizers, five entrees, and three desserts. Assuming the courses are served in that order, and you have one of each, how many different meals can you have?

Steps	Example
1. Place a dash for each event (appetizer, entrée, and dessert.)	___ ___ ___
2. Write the number of possibilities for each event	6 5 3
3. Multiply these for the number of meals possible	6 × 5 × 3 = 90

DRILL	CORE COUNTING METHOD

A company is putting together a team consisting of a developer, a business analyst, and a project manager. If there are 8, 5, and 4 candidates for each position, respectively, and each candidate is eligible for only one position, how many different teams can be formed?

8 × 5 × 4 = 40 × 4 = 160

SOLUTION NEXT PAGE

WORD PROBLEMS	DRILL	CORE COUNTING METHOD

COUNTING METHODS

A company is putting together a team consisting of a developer, a business analyst, and a project manager. If there are 8, 5, and 4 candidates for each position, respectively, and each candidate is eligible for only one position, how many different teams can be formed?

SOLUTION

Steps	Example
1. Place a dash for each event (developer, business analyst, and project manager)	___ ___ ___
2. Write the number of possibilities for each event	$\underline{8}$ $\underline{5}$ $\underline{4}$
3. Multiply these for the number of teams possible	$\underline{8} \times \underline{5} \times \underline{4} = 160$

QUIZ | CORE COUNTING METHOD | WORD PROBLEMS | COUNTING METHODS

All of the license plates issued in a certain state are designated by a combination of 3 letters followed by either 3 numbers or 4 numbers. Which of the following gives the maximum number of different license plates that can be issued with this combination?

(A) $26^3 \times 10^3$
(B) $26^3 \times 10^4$
(C) $26^3 \times 10^5$
(D) $26^3 \times 10^3 \times 11$
(E) $26^3 \times 10^4 \times 11$

SOLUTION NEXT PAGE

WORD PROBLEMS	QUIZ	CORE COUNTING METHOD

COUNTING METHODS

All of the license plates issued in a certain country are designated by a combination of 3 letters followed by either 3 numbers or 4 numbers. Which of the following gives the maximum number of different license plates that can be issued with this combination?

SOLUTION

STEP 1 **Separate the information from the question**

a combination of 3 letters followed by either 3 numbers or 4 numbers

STEP 2 **Turn this into something you can use**

The license plate shows either "ABC – 123" or "ABC – 1234"

STEP 3 **Use this to answer the question**

1. Place a dash for each event (letters and numbers) — — — — — — —

> The last event has 11 possibilities: 0 through 9 (10 possibilities) and blank (11th possibility, for the 3-number code).

2. Write the number of possibilities for each event $\underline{26}\ \underline{26}\ \underline{26}\ \underline{10}\ \underline{10}\ \underline{10}\ \underline{11}$

3. Multiply these to count the number of license plates possible $\underline{26} \times \underline{26} \times \underline{26} \times \underline{10} \times \underline{10} \times \underline{10} \times \underline{11}$ $= 26^3 \times 10^3 \times 11$

Answer is D

There is another way to work this question:

Add the 3-number combinations

$\underline{26}\ \underline{26}\ \underline{26}\ \underline{10}\ \underline{10}\ \underline{10} = 26^3 \times 10^3 = 1{,}000(26^3)$

to the 4-number combinations

$\underline{26}\ \underline{26}\ \underline{26}\ \underline{10}\ \underline{10}\ \underline{10}\ \underline{10} = 26^3 \times 10^4 = 10{,}000(26^3)$

> This works the same as
>
> 2x
> + 3x
> 5x

$$\begin{array}{r} 1{,}000(26^3) \\ +\ 10{,}000(26^3) \\ \hline 11{,}000(26^3) \end{array} = 26^3 \times 10^3 \times 11$$

TOPIC | **FACTORIALS**

WORD PROBLEMS

COUNTING METHODS

The **factorial** refers to the product of all the integers from that number to 1. This is an arithmetic concept used primarily in counting methods.

$5! = 5 \times 4 \times 3 \times 2 \times 1 = 120$

$4! = 4 \times 3 \times 2 \times 1 = 24$

$1! = 1$

$0! = 1$

In the same way that anything to the zero power equals 1, zero factorial equals 1.

LIFELINE 102

When working with factorials, factor them out and cancel.

Factor them out:

$$\frac{10!}{8!} = \frac{10 \times 9 \times 8 \times 7 \times 6 \times 5 \times 4 \times 3 \times 2 \times 1}{8 \times 7 \times 6 \times 5 \times 4 \times 3 \times 2 \times 1}$$

And cancel:

$$\frac{10!}{8!} = \frac{10 \times 9 \cancel{\times 8 \times 7 \times 6 \times 5 \times 4 \times 3 \times 2 \times 1}}{\cancel{8 \times 7 \times 6 \times 5 \times 4 \times 3 \times 2 \times 1}} = 10 \times 9 = 90$$

DRILL | **FACTORIALS**

Simplify

1) $\frac{11!}{9!}$

2) $\frac{101!}{99!}$

SOLUTION NEXT PAGE

DRILL FACTORIALS

Factor the fraction:

1) $\dfrac{11!}{9!} = \dfrac{11 \times 10 \times 9 \times 8 \times 7 \times 6 \times 5 \times 4 \times 3 \times 2 \times 1}{9 \times 8 \times 7 \times 6 \times 5 \times 4 \times 3 \times 2 \times 1}$

Cancel all the like terms:

$$\frac{11!}{9!} = \frac{11 \times 10 \cancel{\times 9 \times 8 \times 7 \times 6 \times 5 \times 4 \times 3 \times 2 \times 1}}{\cancel{9 \times 8 \times 7 \times 6 \times 5 \times 4 \times 3 \times 2 \times 1}} = 11 \times 10 = 110$$

Factor the fraction, but don't write out all the numbers:

2) $\dfrac{101!}{99!} = \dfrac{101 \times 100 \times 99 \ldots 3 \times 2 \times 1}{99 \ldots 3 \times 2 \times 1}$

Cancel all the like terms:

$$\frac{101!}{99!} = \frac{101 \times 100 \cancel{\times 99 \ldots 3 \times 2 \times 1}}{\cancel{99 \ldots 3 \times 2 \times 1}} = 101 \times 100 = 10{,}100$$

DETAIL FACTORIALS

On the GMAT, you only see the single factorial on a positive integer.	$5!$
You don't see the decimal factorial,	$3.2!$
The double factorial,	$4!!$
or the exponent factorial.	$5^{!}$
FYI: The negative factorial is a non-real number.	$-3!$

QUIZ 1 | FACTORIALS

WORD PROBLEMS

COUNTING METHODS

The greatest prime factor of 14! is

(A) 2
(B) 3
(C) 7
(D) 11
(E) 13

SOLUTION NEXT PAGE

WORD PROBLEMS | QUIZ 1 | FACTORIALS

COUNTING METHODS

The greatest prime factor of 14! is

(A) 2
(B) 3
(C) 7
(D) 11
(E) 13

SOLUTION

STEP 1 Separate the information from the question

14!

STEP 2 Turn this into something you can use

The factors of 14! are $14 \times 13 \times 12 \times 11 \ldots 3 \times 2 \times 1$

STEP 3 Use this to answer the question

The greatest **prime** factor is 13.

Answer is E

QUIZ 2 FACTORIALS

What is the value of $\frac{10!}{7!5!}$?

(A) 10
(B) 6
(C) 3
(D) 2
(E) 1

SOLUTION NEXT PAGE

WORD PROBLEMS	QUIZ 2	FACTORALS

COUNTING METHODS

What is the value of $\dfrac{10!}{7!5!}$?

(A) 10
(B) 6
(C) 3
(D) 2
(E) 1

SOLUTION

STEP 1 Separate the information from the question

$$\frac{10!}{7!5!}$$

STEP 2 Turn this into something you can use

$$\frac{10!}{7!5!} = \frac{10\times 9\times 8\times 7\times 6\times 5\times 4\times 3\times 2\times 1}{(7\times 6\times 5\times 4\times 3\times 2\times 1)(5\times 4\times 3\times 2\times 1)}$$

STEP 3 Use this to answer the question

When canceling two factorials from the bottom, cancel the larger one first.

$$\frac{10!}{7!5!} = \frac{10\times 9\times 8\cancel{\times 7\times 6\times 5\times 4\times 3\times 2\times 1}}{(\cancel{7\times 6\times 5\times 4\times 3\times 2\times 1})(5\times 4\times 3\times 2\times 1)}$$

$$= \frac{\overset{\cancel{5}\ 2}{\overset{\backslash /}{\cancel{10}}} \times \overset{3\ \cancel{3}}{\overset{\backslash /}{\cancel{9}}} \times \overset{\cancel{4}\ \cancel{2}}{\overset{\backslash /}{\cancel{8}}}}{(\cancel{5}\times\cancel{4}\times\cancel{3}\times\cancel{2}\times 1)}$$

$$= 2\times 3$$

$$= 6$$

Answer is B

Which is not a factor of 10! ?

(A) 110
(B) 180
(C) 210
(D) 3^4
(E) 2^8

SOLUTION NEXT PAGE

WORD PROBLEMS | QUIZ 3 | FACTORALS

COUNTING METHODS

Which is not a factor of 10! ?

(A) 110
(B) 180
(C) 210
(D) 3^4
(E) 2^8

SOLUTION

STEP 1 Separate the information from the question

10!

STEP 2 Turn this into something you can use

$10 \times 9 \times 8 \times 7 \times 6 \times 5 \times 4 \times 3 \times 2 \times 1$

STEP 3 Use this to answer the question

Check each answer:

Answer choice	Eliminate
$110 = 11 \times 10$	There is no 11 in 10!, so 110 isn't a factor. Keep this answer choice.
$180 = 2 \times 9 \times 10$	2, 9, and 10 are factors of 10! Eliminate this answer choice.
$210 = 3 \times 7 \times 10$	3, 7, and 10 are factors of 10! Eliminate this answer choice.
$3^4 = 3 \times 3 \times 3 \times 3$	9, 6, and 3 are factors of 10! making 3^4 a factor of 10! Eliminate this answer choice.
$2^8 = 2^3 \times 2^2 \times 2 \times 2 \times 2$	10, 8, 6, 4, and 2 are factors of 10! making 2^8 a factor of 10! Eliminate this answer choice.

$9 = 3 \times 3$
$6 = 3 \times 2$
and there is a solo 3, making $3 \times 3 \times 3 \times 3$

$8 = 2^3$
$4 = 2^2$
and 10, 6, and 2 each has a factor of 2

Answer is A

TOPIC ORDERED COUNTING

WORD PROBLEMS

COUNTING METHODS

Ordered counting refers to the number of possible outcomes when the order of events matters. This is also known as the **permutation.**

LIFELINE 103

When the order matters, place a dash for each event, and write the number of possible outcomes for that event.

Example: Suppose you play 7 CDs. There are 7 events:

___ ___ ___ ___ ___ ___ ___

If you play each CD once, there is a descending number of possibilities per event:

$$\underline{7} \times \underline{6} \times \underline{5} \times \underline{4} \times \underline{3} \times \underline{2} \times \underline{1}$$

Which is the same as

$$7!$$

This is how the factorial works with counting methods.

If you only play 4 CDs, this is the number of possibilities:

$$\underline{7} \times \underline{6} \times \underline{5} \times \underline{4}$$

Which is the same as

$$\frac{7!}{3!}$$

= 10 x 9 x 8 = 720

DRILL ORDERED COUNTING

You have 10 MP3s but only time to hear 3 of them. If the order matters, and you don't play any more than once, how many possibilities are there?

SOLUTION NEXT PAGE

WORD PROBLEMS | DRILL | ORDERED COUNTING

COUNTING METHODS

You have 10 MP3s but only time to hear 3 of them. If the order matters, and you don't play any more than once, how many possibilities are there?

Use the core counting method:

SOLUTION

Steps	**Example**
1. Place a dash for each event (MP3)	___ ___ ___
2. Write the number of possibilities for each event	10 9 8
3. Multiply these to count the number of possibilities	$10 \times 9 \times 8 = 720$

Two parents and three children are to take a trip. If the parents sit in the front row of the car, and the children in the back row, in how many ways can the five of them be seated?

(A) 5
(B) 7
(C) 12
(D) 24
(E) 120

SOLUTION NEXT PAGE

WORD PROBLEMS | QUIZ | ORDERED COUNTING

COUNTING METHODS

Two parents and three children are to take a trip. If the parents sit in the front row of the car, and the children in the back row, in how many ways can the five of them be seated?

(A) 5
(B) 7
(C) 12
(D) 24
(E) 120

SOLUTION

STEP 1 Separate the information from the question

Two parents and three children are to take a trip.

The parents sit in the front row, and the children sit in the back row.

STEP 2 Turn this into something you can use

Two people sit in the front, and three sit in the back

STEP 3 Use this to answer the question

In the front row: $\underline{2} \times \underline{1} = 2$

In the back row: $\underline{3} \times \underline{2} \times \underline{1} = 6$

In the car: (# front row) × (# back row)

1. Place a dash for each event (front row and back row) ___ ___

2. Write the number of possibilities for each event $\underline{2} \times \underline{6}$

3. Multiply these to count the number of ways they can be seated $\underline{2} \times \underline{6} = 12$

Answer is C

TOPIC **UNORDERED COUNTING—COMBINATION**

Unordered counting refers to the number of possible outcomes when the order of events doesn't matter. This is also known as the **combination.**

Suppose you have eight people in a group.

One is to attend a meeting:

There are eight possibilities.

Seven are to attend the meeting:

There are still eight possibilities. This is because when choosing seven to go, you actually choose one to stay behind.

Also: If **all** go to the meeting, there is one possibility. If **none** go to the meeting, there is still one possibility.

If most of the people are selected, and you do the math correctly, you'll get a small number.

DETAIL **UNORDERED COUNTING—COMBINATION**

You know the question is **unordered** when it asks how many people can be selected from a group.

If 10 of us are in class, and two will go get lunch, it doesn't matter who goes first or second.

The question is also **unordered** when it asks how many features something can have.

If the car has 7 available options, and you choose two, the order doesn't matter. The satellite radio doesn't come before or after the moon roof.

CONTINUED NEXT PAGE

WORD PROBLEMS

COUNTING METHODS

LIFELINE 104

When the order doesn't matter, use the formula.

The formula: $$\text{\# of possibilities} = \frac{\text{(whole group)!}}{\text{(those taken)!(those left over)!}}$$

1) With eight people in a group, how many possibilities are there if **two** go to a meeting?

Steps	**Example**
Set up the fraction using the formula	$\frac{8!}{2!6!}$
Simplify the fraction	$\frac{8 \times 7 \cancel{\times 6 \times 5 \times 4} \ldots}{2!\, \cancel{6 \times 5 \times 4} \ldots} = \frac{\overset{4}{\cancel{8}} \times 7}{\underset{1}{\cancel{2}} \times 1}$
Multiply	$= 28$

Remember
When cancelling two factorials from the bottom, cancel the larger one first.

2) With eight people in a group, how many possibilities are there if **six** go to a meeting?

Set up the fraction using the formula $\frac{8!}{6!2!}$

And the answer is the same as 1), because the only difference is the order of the denominator, and the order doesn't matter when multiplying.

3) With eight people in a group, how many possibilities are there if **three** go to a meeting?

Set up the fraction using the formula $\frac{8!}{3!5!}$

Simplify the fraction $\frac{8 \times 7 \times 6 \cancel{\times 5 \times 4} \ldots}{3!\, \cancel{5 \times 4} \ldots}$

$$\frac{\overset{4}{\cancel{8}} \times 7 \times \overset{2}{\cancel{6}}}{\cancel{3 \times 2} \times 1}$$

Multiply $= 56$

When the order doesn't matter, use the formula.

4) With eight people in a group, how many possibilities are there if **five** go to a meeting?

Set up the fraction using the formula $\frac{8!}{5!3!}$

And the answer is the same as before : 56

5) Suppose all eight people in the group are going to the meeting: You know that there's only one way for all 8 to attend, but you can also use the formula.

Set up the fraction using the formula $\frac{8!}{8!0!}$

The 8!'s cancel out, leaving this fraction: $\frac{1}{1 \times 0!}$

This is why 0! = 1

And there is one possibility.

QUIZ 1 UNORDERED COUNTING

A car is to be ordered with 3 of the 5 interior options and 2 of the 4 paint options. With these options, how many different combinations are possible?

(A) 16
(B) 30
(C) 60
(D) 90
(E) 120

SOLUTION NEXT PAGE

WORD PROBLEMS | QUIZ 1 | UNORDERED COUNTING

COUNTING METHODS

A car is to be ordered with 3 of the 5 interior options and 2 of the 4 paint options. With these options, how many different combinations are possible?

(A) 16
(B) 30
(C) 60
(D) 90
(E) 120

SOLUTION

STEP 1 Separate the information from the question

3 of the 5 interior options and 2 of the 4 paint options.

STEP 2 Turn this into something you can use

3 of the 5 interior options:

$$\frac{5!}{3!2!} = \frac{5 \times 4 \cancel{\times 3 \times 2 \times 1}}{(\cancel{3 \times 2 \times 1})(2 \times 1)} = \frac{5 \times 4}{2 \times 1} = 10$$

2 of the 4 paint options:

$$\frac{4!}{2!2!} = \frac{4 \times 3 \cancel{\times 2 \times 1}}{(2 \times 1)(\cancel{2 \times 1})} = \frac{4 \times 3}{2 \times 1} = 6$$

STEP 3 Use this to answer the question

Steps	Example
1. Place a dash for each event (car option)	___ ___
2. Write the number of possibilities for each event (interior and color options)	10 6
3. Multiply these to count the number of possibilities	10 × 6 = 60

Answer is C

Two marbles are to be selected from a box. If there are 45 different possible selections of the 2 marbles, how many marbles are in the box?

(A) 10
(B) 12
(C) 15
(D) 18
(E) 20

SOLUTION NEXT PAGE

WORD PROBLEMS | QUIZ 2 | UNORDERED COUNTING

COUNTING METHODS

Two marbles are to be selected from a box. If there are 45 different possible selections of the 2 marbles, how many marbles are in the box?

(A) 10
(B) 12
(C) 15
(D) 18
(E) 20

SOLUTION

STEP 1 Separate the information from the question

Two marbles are to be selected, and there are 45 different possibilities.

STEP 2 Turn this into something you can use

With this, set up the formula:

$$\text{\# of possibilities} = \frac{\text{(whole group)!}}{\text{(those taken)!(those left over)!}}$$

becomes

$$45 = \frac{x!}{2!(x-2)!}$$

STEP 3 Use this to answer the question

When trying answer choices, start with the easiest number, which in this case is 10.

$$45 = \frac{10!}{2!8!} = \frac{10 \times 9}{2}$$

And it works.

Answer is A

Remember
When the math is complicated, plug in the answer choices. For more on this, see Algebra, p.197.

TOPIC PROBABILITY

Probability refers to how likely something is to occur, and it's always a number between zero and one.

A probability of zero means the event will not happen, and a probability of one means it will happen.

For example, if there's a 50% chance of rain, then there's a 1 out of 2 chance. A 0% chance means it won't rain, and a 100% chance means it will rain.

LIFELINE 105 **When finding probability, find the number of possible outcomes.**

$$\text{Probability of an event} = \frac{\#\text{ you want}}{\#\text{ possible}}$$

If you roll one die:

Steps	**Example**
What's the probability of getting a 3?	
1. There is one that you want. 2. There are six possible outcomes.	$\frac{1}{6}$
What's the probability of getting an odd number?	
1. There are three that you want. 2. There are six possible outcomes.	$\frac{3}{6}$
3. Reduce the fraction	$\frac{1}{2}$

DRILL PROBABILITY

If you roll one die:

1) What's the probability of not getting a 3?

2) What's the probability of getting a 3 or a 4?

SOLUTION NEXT PAGE

WORD PROBLEMS | DRILL | PROBABILITY

PROBABILITY

If you roll one die:

SOLUTION

Steps	Example
1) What's the probability of not getting a 3?	
1. There are five that you want. 2. There are six possible outcomes.	$\frac{5}{6}$
2) What's the probability of getting a 3 or a 4?	
1. There are two that you want. 2. There are six possible outcomes.	$\frac{2}{6}$
3. Reduce the fraction	$\frac{1}{3}$

If three married couples are in a room, and two individuals from this group are to be chosen at random, what is the probability that the chosen couple will be married?

(A) $\frac{1}{5}$

(B) $\frac{2}{5}$

(C) $\frac{1}{3}$

(D) $\frac{3}{5}$

(E) $\frac{2}{3}$

SOLUTION NEXT PAGE

WORD PROBLEMS | QUIZ | PROBABILITY

PROBABILITY

If three married couples are in a room, and two individuals from this group are to be chosen at random, what is the probability that the chosen couple will be married?

(A) $\frac{1}{5}$

(B) $\frac{2}{5}$

(C) $\frac{1}{3}$

(D) $\frac{3}{5}$

(E) $\frac{2}{3}$

SOLUTION

STEP 1 | **Separate the information from the question**

three married couples are in a room, and
two individuals from this group are to be chosen

STEP 2 | **Use this to answer the question**

To find the probability:

Steps	**Example**
1. The number of outcomes possible	$\frac{6!}{2!4!} = \frac{6 \times 5}{2 \times 1} = 15$ possible
2. Find the number of outcomes that you want	3 married couples
3. Reduce the fraction	$\frac{3}{15} = \frac{1}{5}$

Answer is A

TOPIC | **WEIGHTED AVERAGES**

A **weighted average** is like a normal average, except that the items are grouped, averaged, and counted. The larger the group, the more it affects (weighs) the average.

LIFELINE 106

With the weighted average, you can find the ratio.

Example: A group of students has an average score of 100, with an average girls' score of 120 and an average boys' score of 90.

You know that there are more boys in the group, because their average of 90 is closer to the overall average of 100.

To find the ratio:

Steps	**Example**
1. Find the distance of each group from the average	Boys: 90 — Average: 100 — Girls: 120 ←10→ ←20→
2. Cross multiply	Boys: 90 — Average: 100 — Girls: 120 10 ↙ ↘ 20 (crossed arrows)
3. Reduce the fraction	$\frac{\text{Boys}}{\text{Girls}} = \frac{20}{10} = \frac{2}{1}$

This isn't actually cross-multiplying, but it helps you remember the step.

And the ratio of boys to girls is 2 : 1.

DRILL | **WEIGHTED AVERAGES**

A university has only juniors and seniors. The average age of the juniors is 20, the average age of the seniors is 25, and the average age of all the students is 22.

1) Which are there more of: juniors or seniors?
2) What is the ratio of juniors to seniors?

SOLUTION NEXT PAGE

WEIGHTED AVERAGES

A university has only juniors and seniors. The average age of the juniors is 20, the average age of the seniors is 25, and the average age of all the students is 22.

1) Which are there more of: juniors or seniors?

2) What is the ratio of juniors to the number of seniors?

SOLUTION

1) Because the average age of 22 is closer to the juniors' average than the seniors' average, there are more juniors than seniors.

2) What is the ratio of juniors to seniors?

To find the ratio:

Steps	**Example**		
1. Find the distance of each group from the average	Juniors	Average	Seniors
	20	22	25
		←2→	←3→
2. Cross multiply	Juniors	Average	Seniors
	20	22	25
		2	3
3. Use the fraction	$\frac{\text{Juniors}}{\text{Seniors}} = \frac{3}{2}$		

And the ratio of juniors to seniors is 3 : 2.

At a certain company that is hiring a director, the average (arithmetic mean) number of years of experience is 9.1 years for the internal applicants and 9.7 years for the external applicants. If the average number of years of experience for both the internal applicants and the external applicants is 9.3 years, what is the ratio of the number of internal applicants to the number of external applicants?

(A) 2 : 1
(B) 3 : 2
(C) 4 : 3
(D) 5 : 4
(E) 6 : 5

SOLUTION NEXT PAGE

WORD PROBLEMS

WEIGHTED AVERAGES

QUIZ | WEIGHTED AVERAGES

At a certain company that is hiring a director, the average (arithmetic mean) number of years of experience is 9.1 years for the internal applicants and 9.7 years for the external applicants. If the average number of years of experience for both the internal applicants and the external applicants is 9.3 years, what is the ratio of the number of internal applicants to the number of external applicants?

SOLUTION

STEP 1 Separate the information from the question

The total average is 9.3

The internal average is 9.1

The external average is 9.7

STEP 2 Turn this into something you can use

1. Find the distance of each group from the average

Internal	Average	External
9.1	9.3	9.7

←.2→ ←.4→

2. Cross multiply

Internal	Average	External
9.1	9.3	9.7

.2 ↙ ↘ .4

3. Adjust the decimals and reduce the fractions.

$$\frac{\text{Internal}}{\text{External}} = \frac{.4}{.2} = \frac{4}{2} = \frac{2}{1}$$

Remember
When dividing decimals, move the decimal points. For more on this, see Arithmetic, p. 47.

STEP 3 Use this to answer the question

The ratio of internal to external applicants is 2 : 1

Answer is A

TOPIC — **SETTING UP EQUATIONS**

Once the equation is set up, solving it is easy.

LIFELINE 107

When an equation has two unknowns, and finding one tells you the other, use x and (total – x).

Example: Sam invested a total of $800. Part of the money yields 5% annual interest, and the rest yields 3% annual interest. If the total annual interest from this investment is $34, how much did Sam invest at 5% and how much at 3%?

Use x for the amount invested at 5%, but don't use y for the amount at 3%. Instead, use $(800 - x)$.

Steps	**Example**
1. Use x's for the numbers you're looking for	x = amount at 5% $(800 - x)$ = amount at 3%
2. Set up the equation	$x(.05) + (800 - x)(.03) = 34$
3. Solve for x	$.05x + (800)(.03) - .03x = 34$ $.02x + 24 = 34$ $.02x = 10$ $x = 500$
4. Find the amounts invested	$500 went to 5% and $300 went to 3%

SETTING UP EQUATIONS

Remember
When multiplying one number with a decimal and one with zeros, move the decimals.

$800 \times .03 \rightarrow 8 \times 3 \rightarrow 24$

Remember
When dividing numbers with decimals, move the decimals.

$\frac{10}{.02} \rightarrow \frac{1000}{2} \rightarrow 500$

For more on these, see Arithmetic, pp. 45 and 47.

DRILL — **SETTING UP EQUATIONS**

Part of a $6,000 fund was invested at 4 percent and the remainder was invested at 6 percent. The total simple annual interest earned on the $6,000 at the end of one year was $320. How much was invested at 4 percent and how much at 6 percent?

SOLUTION NEXT PAGE

WORD PROBLEMS	DRILL	SETTING UP EQUATIONS

SETTING UP EQUATIONS

Part of a \$6,000 fund was invested at 4 percent and the remainder was invested at 6 percent. The total simple annual interest earned on the \$6,000 at the end of one year was \$320. How much was invested at 4 percent and how much at 6 percent?

SOLUTION

Steps	Example
1. Use x's for the numbers you're looking for	x = amount at 4% $(6{,}000 - x)$ = amount at 6%
2. Set up the equation	$x(.04) + (6{,}000 - x)(.06) = 320$
3. Solve for x	$.04x + (6{,}000)(.06) - .06x = 320$ $-.02x + 360 = 320$ $-.02x = -40$ $x = 2{,}000$
4. Find the amounts invested	\$2,000 went to 4% and \$4,000 went to 6%

A professional athlete takes 500 grams of a daily supplement mixture, supplement *A* and supplement *B*. Supplement *A* contains 10 percent amino acids and supplement *B* contains 60 percent amino acids. If this daily mixture provides exactly 100 grams of amino acids, how many grams of supplement *A* are in the mixture?

(A) 100
(B) 200
(C) 300
(D) 400
(E) 450

SOLUTION NEXT PAGE

WORD PROBLEMS | QUIZ | SETTING UP EQUATIONS

SETTING UP EQUATIONS

A professional athlete takes 500 grams of a daily supplement mixture, supplement *A* and supplement *B*. Supplement *A* contains 10 percent amino acids and supplement *B* contains 60 percent amino acids. If this daily mixture provides exactly 100 grams of amino acids, how many grams of supplement *A* are in the mixture?

SOLUTION

STEP 1 Separate the information from the question

500 grams per day
Supplement *A* has 10% aminos
Supplement *B* has 60% aminos
100 grams of aminos total

STEP 2 Turn this into something you can use

Steps	Example
1. Use x's for the numbers you're looking for	x = amount at 10% $(500 - x)$ = amount at 60%

STEP 3 Use this to answer the question

2. Set up the equation	$x(.1) + (500 - x)(.6) = 100$
3. Solve for x	$.1x + (500)(.6) - .6x = 100$ $-.5x + 300 = 100$ $-.5x = -200$ $x = 400$

Answer is D

HOMEWORK

WORD PROBLEMS

13th Editon *GMAT Review Book,* pp. 152 -185, question numbers

GMAT REVIEW

ANSWERS NEXT PAGE

Work Process
23 ____
81 ____
86 ____
103 ____
139 ____
168 ____

Rate / Distance
79 ____
162 ____
207 ____

Conversions
38 ____
59 ____
63 ____
65 ____
70 ____

Counting Methods
39 ____
128 ____
133 ____
148 ____
228

Factorials
116 ____

Set Distribution
178 ____
201 ____
222 ____

Sets in Tables
33 ____
96 ____
107 ____
136 ____
186 ____

Setting Up Equations
88 ____
169 ____
224 ____

Weighted Averages
208 ____

Probability
10 ____
68 ____
173 ____
193 ____
215 ____

WORD PROBLEMS	HOMEWORK
GMAT REVIEW	**13th Editon** *GMAT Review Book,* pp. 152 -185, question numbers
ANSWERS	

Work Process
23 ____B
81 ____E
86 ____B
103 ____D
139 ____C
168 ____B

Rate / Distance
79 ____A
162 ____E
207 ____E

Conversions
38 ____A
59 ____E
63 ____A
65 ____B
70 ____D

Counting Methods
39 ____B
128 ____B
133 ____C
148 ____C
228 ____C

Factorials
116 ____C

Set Distribution
178 ____D
201 ____D
222 ____D

Sets in Tables
33 ____B
96 ____B
107 ____C
136 ____B
186 ____E

Setting Up Equations
88 ____C
169 ____D
224 ____B

Weighted Averages
208 ____E

Probability
10 ____A
68 ____B
173 ____E
193 ____D
215 ____E

Data Sufficiency

Data Sufficiency

Lesson 5

About half of the Quantiative questions are in Data Sufficiency format.

For GMAT study purposes, this question type has been separated into its own lesson to focus on specific strategies and specific math concepts that often appear in this format.

The scope of the math is the same as in the preceding lessons, so this lesson also serves as a review.

TOPIC	QUESTIONS & ANSWER CHOICES	DATA SUFFICIENCY
		STRATEGIES & TRAPS

The Data Sufficiency format asks a question and provides two relevant pieces of information, labeled (1) and (2).

For example:

Is $2x - 3$ equal to 1?

(1) $x > 1$

(2) $2x + 3 < 9$

You determine whether the information, or data, is sufficient to answer the question.

The answer choices never change:

(A) Statement (1) ALONE is sufficient, but statement (2) alone is not sufficient.

(B) Statement (2) ALONE is sufficient, but statement (1) alone is not sufficient.

(C) BOTH statements TOGETHER are sufficient, but NEITHER statement ALONE is sufficient.

(D) EACH statement ALONE is sufficient.

(E) Statements (1) and (2) TOGETHER are NOT sufficient.

On test day, you don't want to read and interpret these answer choices each time you see them.

LIFELINE 108

When answering a data sufficiency question, use these answer choices.

To answer the question, you need:

(A) (1) alone
(B) (2) alone
(C) Need both
(D) Either one
(E) Can't answer

DATA SUFFICIENCY

STRATEGIES & TRAPS

LIFELINE 109

When checking the statements, pay attention to how they fit together.

Three things about the statements are always true:

1. The statements never disagree:

In this example on the left, each statement alone answers the question.

However, because x can't equal both 2 and 3, this type of contradiction won't appear on the GMAT.

> **You will not see:**
>
> What is the value of x?
>
> (1) $x = 2$
>
> (2) $x = 3$

2. They never repeat information that's in the question.

Because this question on the left already tells you that x is positive and an integer, that information will not be repeated in the statements.

> **You will not see:**
>
> If x is a positive integer, is $3x$ an even integer?
>
> (1) x is an integer
>
> (2) x is positive

3. However, the statements could both tell you the same thing.

In this example on the left, each statements tells you the value of x, so the answer is (D) Either one.

> **You could see:**
>
> What is the value of x?
>
> (1) $x = 3$
>
> (2) $x + 2 = 5$

LIFELINE 110

When the statements repeat each other, the only possible answers are (D) Either one or (E) Can't answer.

Another example:

What is the value of y?

(1) $x + y = 90$

(2) $x = 90 - y$

These statements tell you the same thing, but they don't answer the question, so the answer is (E) Can't answer.

TOPIC	WORKING A DATA SUFFICIENCY QUESTION

These core strategies get you through the questions.

LIFELINE 111 **When working a Data Sufficiency question, cover up the statements and simplify the question.**

Example:

Is $2x - 3$ equal to 1?

(Statements are covered)

Steps	**Example**
To simplify	"Is $2x - 3$ equal to 1?"
Add 3 to both sides:	"Is $2x$ equal to 4?"
Divide both sides by 2	"Is x equal to 2?"

The question really asks, "Is x equal to 2?"

This is simpler than the original question.

Evaluate the statements separately: 1

On your scratch paper, write down 1 and 2. 2

Look at Statement (1).

(1) $x > 1$

If x is greater than 1, it **could** be equal to 2, but it could also be equal to 3.

Statement (1) isn't sufficient, so cross off the 1 on your scratch paper. ~~1~~

DATA SUFFICIENCY | **TOPIC** | **WORKING A DATA SUFFICIENCY QUESTION — Continued**

STRATEGIES & TRAPS

Now look at Statement (2).

(2) $2x + 3 < 9$

LIFELINE 112

When evaluating a complicated statement, simplify it first.

Steps	**Example**
To simplify	$2x + 3 < 9$
Subtract 3 from both sides	$2x < 6$
Divide both sides by 2	$x < 3$

If x is less than 3, it **could** be equal to 2, or it could be equal to 1.

"Could be" doesn't work for the GMAT.
Cross off Statement (2). ~~2~~

Now that both statements are crossed off:

The only possible answer choices are
(C) Need both; and (E) Can't answer.

Try both statements together to answer the question.

Is $x = 2$?

$x > 1$ and $x < 3$

So is it (C) or (E)?

You would think, "Yes, 2 is between 1 and 3."
The common wrong answer is (C) Need both.

Answer is (E) Can't answer

The reason is that you can't assume x is an integer.
x could be equal to 1.5 or 2.5.

Devious? Yes. Like the GMAT? Absolutely.
You can safely fall for the trap here in this lesson, but not on the actual test. Soon you will recognize most of the common GMAT Data Sufficiency traps.

TOPIC	GUESSING STRATEGY	DATA SUFFICIENCY

STRATEGIES & TRAPS

If you resolve one statement, but not the other, you can eliminate some of the answer choices

LIFELINE 113 **If you have to guess, first narrow the possibilities.**

If 1 is circled, but you can't tell about 2, ①

Choose (A) (1) alone 2 ?

→ (A) (1) alone
(B) (2) alone
(C) Need both
(D) Either one
(E) Can't answer

If 2 is circled, but you can't tell about 1, 1 ?

Choose (B) (2) alone ②

If one statement is crossed off and you can't tell about the other, ~~1~~ 2 ? —OR— 1 ? ~~2~~

or if both are crossed off, but you still can't tell whether you can solve it, ~~1~~ ~~2~~

Choose (E) Can't answer

DETAIL GUESSING STRATEGY

For each situation, more than one answer choice is possible.

For example, if you have this: ~~1~~ 2 ?

Then the correct answer could be B, C, or E. But you can only choose one answer, so why memorize three possibilities?

Now that you have the strategies, practice with the following 28 Data Sufficiency questions based on math concepts commonly appearing in this format.

Each math concept is covered in the previous lessons. In this sense, this chapter also serves as a math review.

For each question, the relevant Life-line and page number is included for your reference.

QUIZ	DIVISIBILITY	DATA SUFFICIENCY
LIFELINE 15	**When dividing a number, use number properties instead of long division.**	ARITHMETIC

If x is a prime number, what is the least common multiple of x, 2, and 3 ?

(1) x is a factor of 125

(2) The least common multiple of x and 3 is 15

To answer the question, you need:

(A) (1) alone
(B) (2) alone
(C) Both together
(D) Either one
(E) Can't answer

Lifeline 15 is on p. 23

1

2

Write 1 and 2 on your scratch paper.

SOLUTION NEXT PAGE

DATA SUFFICIENCY | QUIZ | FACTORS AND MULTIPLES

ARITHMETIC

If x is a prime number, what is the least common multiple of x, 2, and 3 ?

SOLUTION

STEP 1 | **What is the question really asking or what do you need**

The least common multiple of three different prime numbers, such as 2, 3, and 5, is the product of those numbers: $2 \times 3 \times 5$.

For the least common multiple of 2, 3, and x, you need to know what x is.

The question is really asking, "What is the value of x?"

STEP 2 | **Simplify one statement to determine what it tells you**

Remember
A number ending in 5 is divisible by 5. For more on number properties, see Arithmetic, p. 23.

(1) x is a factor of 125

The factors of 125 are 1, 5, 25, and 125.

You know x is prime from the question, and 5 is the only prime number in this list, so $x = 5$.

Statement (1) is sufficient. Circle the 1.

(1)

STEP 3 | **Simplify the other statement to determine what it tells you**

Statement (2) could also mean that $x = 15$, 30, or any multiple of 15. However, because x is prime, it has to be a factor of 15, and it couldn't be 3.

(2) The least common multiple of x and 3 is 15

In other words, $(3)(x) = 15$, making $x = 5$.

Statement (2) is sufficient. Circle the 2.

(2)

1 and 2 are both circled.

Answer is (D) Either one

QUIZ	ODD TIMES EVEN	DATA SUFFICIENCY
LIFELINE 11	**When *x* is odd or even, use real odd or even numbers, such as 2 or 3.**	ARITHMETIC

If a and b are positive integers, is the product ab even?

(1) $3a + 2b$ is even

(2) $4a - 5b$ is even

To answer the question, you need:

(A) (1) alone
(B) (2) alone
(C) Both together
(D) Either one
(E) Can't answer

Lifeline 11 is on p. 9

1

2

Write 1 and 2 on your scratch paper.

SOLUTION NEXT PAGE

DATA SUFFICIENCY | QUIZ | ODD TIMES EVEN

ARITHMETIC

If a and b are positive integers, is the product ab even?

SOLUTION

STEP 1 | **What is the question really asking or what do you need**

The product ab is even if either a or b is even, or both.

For example, both 2×2 and 2×3 are even.

The question is really asking, "Is either a or b even?"

STEP 2 | **Simplify one statement to determine what it tells you**

(1) $3a + 2b$ = even.

Reason	**Example**
You know $2b$ is even because 2 times any integer is even.	2×5 is even
If $3a$ plus an even number is even, $3a$ must also be even.	$4 + 4$ is even, while $4 + 3$ is odd
If $3a$ is even, a is even, because 3 times an odd number is odd.	3×4 is even, while 3×5 is odd

This makes a even

(1) Statement (1) is sufficient. Circle the 1.

STEP 3 | **Simplify the other statement to determine what it tells you**

(2) $4a - 5b$ is even

You know $4a$ is even because 4 times any integer is even.	4×5 is even
If an even number minus $5b$ is even, $5b$ must also be even.	$6 - 4$ is even, while $6 - 3$ is odd
If $5b$ is even, b is even, because 5 times an odd number is odd.	3×2 is even, while 3×3 is odd

This makes b even

(2) Statement (2) sufficient. Circle the 2.

Answer is (D) Either one

 thinktankSCHOLASTIC

QUIZ	REMAINDERS FROM DIVIDING	DATA SUFFICIENCY
LIFELINE 12	**When x divided by a number has a certain remainder, find some possible values of x.**	ARITHMETIC

If r is the remainder when the positive integer n is divided by 5, what is the value of r?

(1) When n is divided by 10, the remainder is 2

(2) n is a multiple of 6

To answer the question, you need:

(A) (1) alone
(B) (2) alone
(C) Both together
(D) Either one
(E) Can't answer

1

2

Lifeline 12 is on p. 11

SOLUTION NEXT PAGE

DATA SUFFICIENCY | **QUIZ** | **REMAINDERS FROM DIVIDING**

ARITHMETIC

If r is the remainder when the positive integer n is divided by 5, what is the value of r?

SOLUTION

STEP 1 What is the question really asking or what do you need

$$\frac{n}{5} = (\#)\ r\ \underline{?}$$

Or, "what is the remainder when n is divided by 5? "

STEP 2 Simplify one statement to determine what it tells you

(1) When n is divided by 10, the remainder is 2

n could equal 12, 22, or 32.

To answer the question, divide these by 5:

$\frac{12}{5}$, $\frac{22}{5}$, and $\frac{32}{5}$ each has the same remainder of 2

(1) Statement (1) is sufficient. Circle the 1 on your scratch paper.

STEP 3 Simplify the other statement to determine what it tells you

(2) n is a multiple of 6

n could equal 6, 12, or 18.

To answer the question, divide these by 5.

Each has a different remainder.

~~2~~ Statement (2) not sufficient. Cross off the 2 on your scratch paper.

Answer is (A) 1 alone

With both statements together, you might determine that n equals 12. However, the question isn't asking for the value of n, so C isn't the correct answer.

Also, there are other numbers, such as 72, that are both divisible by 6 and have a remainder of 2 when divided by 10.

QUIZ	FRACTIONS	DATA SUFFICIENCY
LIFELINE 22	**When working with fractions, cancel what you can.**	ARITHMETIC

If n and d are positive integers, is $\frac{n}{d}$ an integer?

(1) Each factor of d is also a factor of n

(2) No two factors of d are equal

To answer the question, you need:

(A) (1) alone
(B) (2) alone
(C) Both together
(D) Either one
(E) Can't answer

1

2

Lifeline 22 is on p. 37

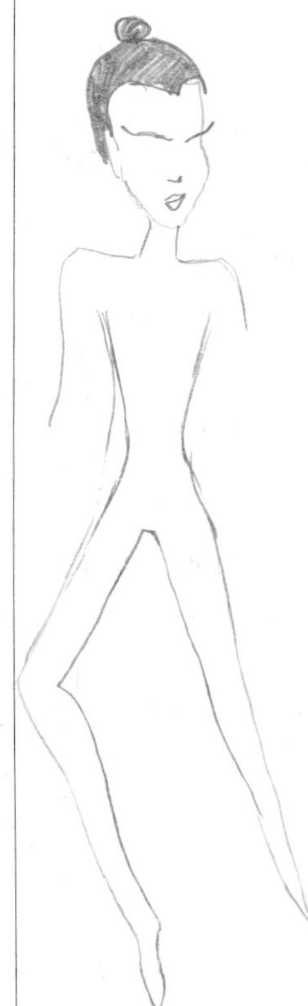

SOLUTION NEXT PAGE

DATA SUFFICIENCY | QUIZ | FRACTIONS

ARITHMETIC

If n and d are positive integers, is $\frac{n}{d}$ an integer?

SOLUTION

Remember
A fraction becomes an integer when all the numbers on the bottom cancel out. For more on this, see Arithmetic, p. 22.

STEP 1 What is the question really asking or what do you need

"Do all the factors of d cancel out with the factors of n?"

For example, $\frac{n}{d} = \frac{\cancel{2 \times 3} \times 5}{\cancel{2 \times 3}}$

STEP 2 Simplify one statement to determine what it tells you

(1) Each factor of d is also a factor of n

In other words, you could have this, where d cancels: $\frac{n}{d} = \frac{2 \times 3}{2 \times 3}$

or this, because the factors of d could be repeated: $\frac{n}{d} = \frac{2 \times 3}{2 \times 3 \times 3}$

~~1~~ Statement (1) isn't sufficient. Cross off 1 on your scratch paper.

STEP 3 Simplify the other statement to determine what it tells you

(2) No two factors of d are equal

Without information from statement (1) you could have this: $\frac{n}{d} = \frac{2 \times 3}{2 \times 3 \times 5}$

or this: $\frac{n}{d} = \frac{2 \times 3 \times 5}{2 \times 3 \times 5}$

~~2~~ Statement (2) isn't sufficient. Cross off 2 on your scratch paper.

STEP 4 Both 1 and 2 are crossed off, so check the statements together

If each factor of d is also a factor of n, and no two factors of d are equal, each factor of d will cancel with a factor of n.

For example: $\frac{n}{d} = \frac{2 \times 3 \times 5 \times 7}{2 \times 3 \times 5}$

Both statements together are needed.

Answer is (C) Need both

QUIZ | **FRACTIONS** | **DATA SUFFICIENCY**

LIFELINE 22

When working with fractions, cancel what you can.

ARITHMETIC

What is the value of $\dfrac{an + ad}{n + d}$?

(1) $a = 2$

(2) $n + d = 3$

To answer the question, you need:

(A) (1) alone
(B) (2) alone
(C) Both together
(D) Either one
(E) Can't answer

Lifeline 22 is on p. 37

1

2

SOLUTION NEXT PAGE

DATA SUFFICIENCY | QUIZ | FRACTIONS

ARITHMETIC

What is the value of $\dfrac{an + ad}{n + d}$?

SOLUTION

STEP 1 **What is the question really asking or what do you need**

Start with the fraction. $\dfrac{an + ad}{n + d}$

Factor the *a*'s from the numerator: $\dfrac{a(n + d)}{n + d}$

Cancel each $(n + d)$ $\dfrac{a\cancel{(n + d)}}{\cancel{n + d}}$

And the question really asks: a
"What is the value of a ?"

STEP 2 **Simplify one statement to determine what it tells you**

(1) $a = 2$

This answers the question.

1 Statement (1) is sufficient. Circle the 1 on your scratch paper.

STEP 3 **Simplify the other statement to determine what it tells you**

(2) $n + d = 3$

This doesn't answer the question, but it is a trap for those who don't simplify the question first.

~~2~~ Statement (2) isn't sufficient. Cross off the 2 on your scratch paper.

Answer is (A) 1 alone

QUIZ | FRACTIONS | DATA SUFFICIENCY

LIFELINE 22

When working with fractions, cancel what you can.

ARITHMETIC

Lifeline 22 is on p. 37

What is the value of $\frac{x + 2y}{x + y}$?

(1) $x + y = 5$

(2) $\frac{y}{x + y} = 2$

To answer the question, you need:

(A) (1) Alone
(B) (2) Alone
(C) Both together
(D) Either one
(E) Can't answer

1

2

SOLUTION NEXT PAGE

DATA SUFFICIENCY | QUIZ | FRACTIONS

ARITHMETIC

What is the value of $\frac{x + 2y}{x + y}$?

SOLUTION

STEP 1 What is the question really asking or what do you need

Start with the fraction. $\frac{x + 2y}{x + y}$

Subtract the parts of the fraction that cancel. $\frac{x + 2y}{x + y} = \frac{y}{x + y} + \frac{x + y}{x + y}$

This works the same way that:

$\frac{a}{2} + \frac{b}{2} = \frac{a + b}{2}$

And the $(x + y)$'s cancel from the fraction on the right. $\frac{y}{x + y} + \frac{\cancel{x + y}}{\cancel{x + y}}$

And the question really asks, "What is the value of this fraction?" $\frac{y}{x + y}$

STEP 2 Simplify one statement to determine what it tells you

(1) $x + y = 5$

This doesn't answer the question, but it is a trap for those who don't simplify the question first.

~~1~~ Statement (1) isn't sufficient. Cross off the 1 on your scratch paper.

STEP 3 Simplify the other statement to determine what it tells you

(2) $\frac{y}{x + y} = 2$

This answers the question.

(2) Statement (2) is sufficient. Circle the 2 on your scratch paper.

Answer is (B) 2 alone

QUIZ	EXPONENTS	DATA SUFFICIENCY
LIFELINE 41	**When dividing numbers with exponents, subtract the exponents.**	ARITHMETIC

If n is a positive integer, and $d = \dfrac{3.4 \times 10^n}{1.7 \times 10^3}$, what is the value of d ?

(1) $10 > d > 100$

(2) $n - 3 > 0$

To answer the question, you need:

(A) (1) alone
(B) (2) alone
(C) Both together
(D) Either one
(E) Can't answer

Lifeline 41 is on p. 86

1

2

SOLUTION NEXT PAGE

DATA SUFFICIENCY | QUIZ | EXPONENTS

ARITHMETIC

If n is a positive integer, and $d = \dfrac{3.4 \times 10^n}{1.7 \times 10^3}$, what is the value of d ?

SOLUTION

STEP 1 **What is the question really asking or what do you need**

Start with the fraction $d = \dfrac{3.4 \times 10^n}{1.7 \times 10^3}$

Separate the fraction into manageable pieces $d = \dfrac{3.4}{1.7} \times \dfrac{10^n}{10^3}$

This works the same way that: $\dfrac{2}{3} \times \dfrac{a}{b} = \dfrac{2a}{3b}$

Simplify the fraction as $\dfrac{3.4}{1.7} = 2$ and $\dfrac{10^n}{10^3} = 10^{n-3}$

n is a positive integer $n = 1, 2, 3, ...$

giving $d = 2 \times 10^{n-3}$ $d = 2 \times 10^{n-3}$

some possible values:

$$d = 2 \times 10^{-2} = .02$$
$$2 \times 10^{-1} = .2$$
$$2 \times 10^{0} = 2$$
$$2 \times 10^{1} = 20$$
$$2 \times 10^{2} = 200$$

STEP 2 **Simplify one statement to determine what it tells you**

(1) $10 > d > 100$

From the possible values of d listed above, the only value between 10 and 100 is 20.

(1) Statement (1) is sufficient. Circle the 1 on your scratch paper.

STEP 3 **Simplify the other statement to determine what it tells you**

(2) $n - 3 > 0$

Making the exponent on the 10 equal to possibly 1, 2, or 3, and d equal to possibly 20, 200, or 2,000.

~~2~~ Statement (2) is not sufficient. Cross off the 2 on your scratch paper.

Answer is (A) 1 alone

QUIZ	COMBINED RATIOS	DATA SUFFICIENCY
LIFELINE 38	**When combining ratios, stack the part that they have in common.**	ARITHMETIC

Lifeline 38 is on p. 81

At a certain motorcycle dealership, the ratio of the number of Brand X motorcycles to the number of Brand Y motorcycles was 3 to 2. What was the number of Brand Z motorcycles at the dealership?

(1) At the dealership, the ratio of the number of Brand Y motorcycles to the number of Brand Z motorcycles was 3 to 4.

(2) The number of Brand Y motorcycles at the dealership was 12.

To answer the question, you need:

(A) (1) Alone
(B) (2) Alone
(C) Both together
(D) Either one
(E) Can't answer

1

2

SOLUTION NEXT PAGE

DATA SUFFICIENCY | QUIZ | ARITHMETIC / COMBINED RATIOS

QUESTION TYPES

At a certain motorcycle dealership, the ratio of the number of Brand X motorcycles to the number of Brand Y motorcycles was 3 to 2. What was the number of Brand Z motorcycles at the dealership?

SOLUTION

STEP 1 | **What is the question really asking or what do you need**

Given a ratio with three items, this is a combined ratio question. Set up what you can from the question.

X	:	Y	:	Z
3	:	2		

STEP 2 | **Simplify one statement to determine what it tells you**

(1) At the dealership, the ratio of the number of Brand Y motorcycles to the number of Brand Z motorcycles was 3 to 4.

For more on combined ratios, see Arithmetic p. 81.

This allows you to complete the combined ratio.

However, it's still only a ratio. The actual numbers can be any multiple of these.

X	:	Y	:	Z
3	:	2	:	
	:	3	:	4
9	:	6	:	8

~~1~~ Statement (1) is not sufficient. Cross off the 1.

STEP 3 | **Simplify the other statement to determine what it tells you**

(2) The number of Brand Y motorcycles at the dealership was 12.

By itself, without using information from Statement (1) the number of Brand Y motorcycles has nothing to do with the number of Brand Z motorcycles.

~~2~~ Statement (2) is not sufficient. Cross off the 2.

STEP 4 | **Both 1 and 2 are crossed off, so check the statements together**

Given the combined ratio, the numbers of Brands X, Y, and Z motorcycles could be any numbers that fit the ratio:

X	:	Y	:	Z
9	:	6	:	8
18	:	12	:	16
27	:	18	:	24

However, the number of Brand Y motorcycles is 12, making the number of Brand Z motorcycles 16.

Both statements taken together are sufficient.

Answer is (C) Both together

QUIZ	EXPONENTS	DATA SUFFICIENCY
LIFELINE 44	**When multiplying or dividing numbers with the same exponent, keep the exponent.**	ARITHMETIC

If n is an integer, is $2^n \times 3^5 > 12^5$?

(1) $n > 10$

(2) $n < 12$

To answer the question, you need:

(A) (1) alone
(B) (2) alone
(C) Both together
(D) Either one
(E) Can't answer

Lifeline 44 is on p. 91

1

2

SOLUTION NEXT PAGE

DATA SUFFICIENCY | QUIZ | EXPONENTS

ARITHMETIC

If n is an integer, is $2^n \times 3^5 > 12^5$?

SOLUTION

STEP 1 What is the question really asking or what do you need

Simplify the question.

Remember
When working with an inequality, for the most part, treat it like an equal sign.

For more on this, see Algebra, p.137.

Steps	**Example**
Divide both sides by 3^5	$\frac{12^5}{3^5} = 4^5$
Now the question is simpler:	Is $2^n > 4^5$?
Convert 4^5 to 2^{10}	$4^5 = 2^{10}$
Now the question is even simpler:	Is $2^n > 2^{10}$?
and it really asks,	"Is $n > 10$?"

STEP 2 Simplify one statement to determine what it tells you

(1) $n > 10$

This statement answers the question.

(1) Statement (1) is sufficient. Circle the 1.

STEP 3 Simplify the other statement to determine what it tells you

(2) $n < 12$

n could be 11 or 9

~~2~~ Statement (2) is not sufficient. Cross off the 2.

Answer is A 1 alone

Taken together, you know $n = 11$. However, the question doesn't ask for the value of n.

thinktankSCHOLASTIC

QUIZ	RADICALS	DATA SUFFICIENCY
LIFELINE 49	**When completing a radical, factor it out**	ARITHMETIC

Lifeline 49 is on p. 109

If y and n are positive integers, is $180y$ the square of an integer?

(1) $y = 5^n$

(2) n is odd

To answer the question, you need:

(A) (1) alone
(B) (2) alone
(C) Both together
(D) Either one
(E) Can't answer

1

2

SOLUTION NEXT PAGE

DATA SUFFICIENCY	QUIZ	RADICALS

ARITHMETIC

SOLUTION

STEP 1 **What is the question really asking or what do you need**

If $180y$ is the square of an integer, it can be written like this: $180y = \text{int}^2$

Square root both sides $\sqrt{180y} = \text{int}$

And factor the 180 to see what y has to be to complete the square root.

A square root needs numbers in pairs. The 3's and 2's are in pairs, so they're fine.

The 5, however, doesn't have a match.

180
/ \
18 10
/ | \ / \
3 3 2 2 5

2's and 3's are paired

STEP 2 **The question really asks, "Does y provide the missing 5?"**

Because n is an integer, statement (1) tells you that y is an exponent of 5.

(1) $y = 5^n$ y could equal 5
5×5
$5 \times 5 \times 5$
and so on.

y provides the missing 5, but it also provides extra 5's. That's OK, as long as all the extra fives are also in pairs, which is the case if the exponent, n, is odd.

However, if $y = 5^4$, there will be an extra y.

For example, if $y = 5^3$, all the numbers are in pairs, and $180y$ is the square of an integer

$y = 5 \times 5 \times 5$

Pairs of 5's

$$180y = \underbrace{2 \times 2 \times 3 \times 3 \times 5}_{180} \times \underbrace{5 \times 5 \times 5}_{y}$$

Statement (1) doesn't tell you enough about n, so you don't know whether all the 5's in $180y$ are in pairs.

~~1~~ Statement (1) is not sufficient. Cross off the 1.

STEP 3 **Simplify the other statement to determine what it tells you**

(2) n is odd By itself, you don't know how n fits into the equation.

~~2~~ Statement (2) is not sufficient. Cross off the 2.

STEP 4 **Both 1 and 2 are crossed off, so check the statements together**

From Statement (1), you know that if n is odd, all the 5's (and thus all the factors) are in pairs, making $\sqrt{180y}$ an integer.

Answer is (C) Both together

QUIZ | ARITHMETIC MEAN | DATA SUFFICIENCY

LIFELINE 50

When working with the average, set up the equation with what you have.

ARITHMETIC

If the median of the five numbers *x*, 2, 1, 11, and 6 is equal to the average (arithmetic mean) of the five numbers, what is the value of *x* ?

(1) *x* is the median of the five numbers

(2) $2 < x < 6$

To answer the question, you need:

(A) (1) alone
(B) (2) alone
(C) Both together
(D) Either one
(E) Can't answer

1

2

Lifeline 50 is on p. 115

SOLUTION NEXT PAGE

DATA SUFFICIENCY	QUIZ	ARITHMETIC MEAN

ARITHMETIC

If the median of the five numbers x, 2, 1, 11, and 6 is equal to the average (arithmetic mean) of the five numbers, what is the value of x ?

SOLUTION

STEP 1 What is the question asking or what do you need

Because the average is equal to the median,

$$\frac{x + 2 + 1 + 11 + 6}{5} = median$$

$$\frac{20 + x}{5} = median$$

Find the median, and you can solve for x.

The question is really asking, "What is the median?"

STEP 2 Simplify one statement to determine what it tells you

(1) x is the median of the five numbers

Meaning you can substitute x for the word "median" in the equation

$$\frac{20 + x}{5} = x$$

Now, you can solve for x.

1 Statement (1) alone is sufficient. Circle the 1.

STEP 3 Simplify the other statement to determine what it tells you

(2) $2 < x < 6$

Meaning if you place the numbers in order, they look like this:

1, 2, x, 6, 11

Making x the median, and you can set up the equation:

$$\frac{20 + x}{5} = x$$

Statement (2) tells you the same thing as Statement (1).

2 Statement (2) alone is sufficient. Circle the 2.

> **Remember**
> When working with the median, put the numbers in order.
>
> For more on this, see Arithmetic, p.119.

Answer is (D) Either one

To actually solve for x, multiply both sides by 5

$$\frac{20 + x}{5} = x$$

$$20 + x = 5x$$

Subtract x from both sides

$$20 = 4x$$

And divide both sides by 4

$$5 = x$$

QUIZ	NUMBER LINE	DATA SUFFICIENCY
LIFELINE 56	**When working with the number line, draw the number line.**	ALGEBRA

If *A*, *B*, and *C* are points on the number line, but not necessarily in that order, is *B* between *A* and *C*?

(1) The distance between *A* and *B* is the same as the distance between *A* and *C*

(2) *A* is to the left of *B*

To answer the question, you need:

(A) (1) alone
(B) (2) alone
(C) Both together
(D) Either one
(E) Can't answer

Lifeline 56 is on p. 133

1

2

SOLUTION NEXT PAGE

DATA SUFFICIENCY | **QUIZ** | **NUMBER LINE**

ALGEBRA

If *A*, *B*, and *C* are points on the number line, but not necessarily in that order, is *B* between *A* and *C*?

SOLUTION

STEP 1 What is the question really asking or what do you need

The question is asking whether one of these is the case:

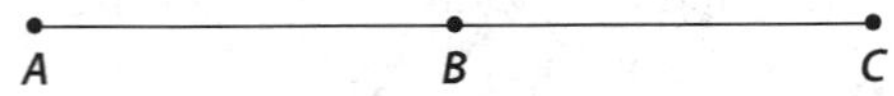

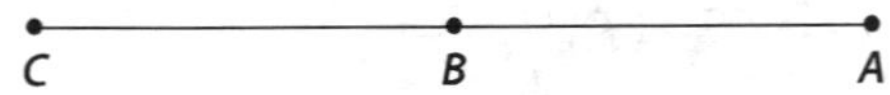

STEP 2 Simplify one statement to determine what it tells you

> When drawing a number line with points labeled with letters (*A, B, C*, etc.), never assume the letters are in alphabetical order.

(1) The distance between *A* and *B* is the same as the distance between *A* and *C*

Either *A* and *C* are on opposite sides of *B*

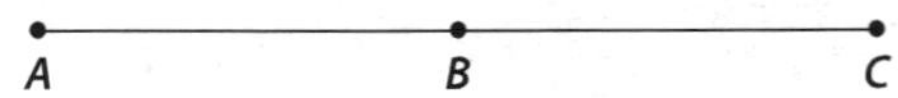

or *A* and *C* are the same point.

~~1~~ Statement (1) is not sufficient. Cross off the 1.

STEP 3 Simplify the other statement to determine what it tells you

(2) *A* is to the left of *B*

This doesn't tell you whether *B* is between *A* and *C*.

~~2~~ Statement (2) is not sufficient. Cross off the 2.

STEP 4 Both 1 and 2 are crossed off, so check the statements together

With *A* and *C* placed equally apart from *B*, and *A* to the left of *B*,

you still don't know whether the line looks like this

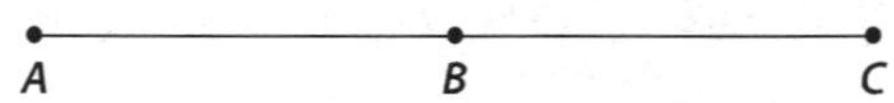

or this, if *A* and *C* are the same point

Answer is (E) Can't answer

QUIZ	INEQUALITIES	DATA SUFFICIENCY
LIFELINE 58	**When working with an inequality, for the most part, treat it like an equal sign.**	ALGEBRA

If $xy > b$, is x positive?

(1) $b = 3y$

(2) y is positive

To answer the question, you need

(A) 1 alone
(B) 2 alone
(C) 1 and 2 together
(D) Either one alone
(E) Can't answer

1

2

Lifeline 58 is on p. 137

SOLUTION NEXT PAGE

DATA SUFFICIENCY | QUIZ | INEQUALITIES

INEQUALITIES

If $xy > b$, is x positive?

SOLUTION

STEP 1 What is the question really asking or what do you need

The question really asks, "Are y and b positive or negative?"

STEP 2 Simplify one statement to determine what it tells you

Remember
When dividing by an unknown, such as y, you don't know whether it's positive or negative.

For more on this, see Algebra, p. 137.

(1) $b = 3y$

If $xy > b$ and $b = 3y$, then combine the equations: $xy > 3y$

Divide both sides by y for $x > 3$

However, you don't know whether to switch the sign.

~~1~~ Statement (1) is not sufficient. Cross off the 1.

STEP 3 Simplify the other statement to determine what it tells you

(2) y is positive

With $xy > b$ and y being positive, b and x could be positive or negative.

~~2~~ Statement (2) is not sufficient. Cross off the 2.

STEP 4 Both 1 and 2 are crossed off, so check the statements together

Because both Statements 1 and 2 are crossed off,

From $xy > 3y$, you know $x > 3$.

Since $x > 3$, it must be positive.

Both statements together are sufficient. The answer is C.

Answer is (C) Both together

QUIZ | COORDINATE GEOMETRY | DATA SUFFICIENCY

LIFELINE 60

To find the equation of a line, find the y-intercept and the slope.

ALGEBRA

Lifeline 60 is on p. 143

In the rectangular coordinate system shown, does the line ℓ (not shown) intersect Region III?

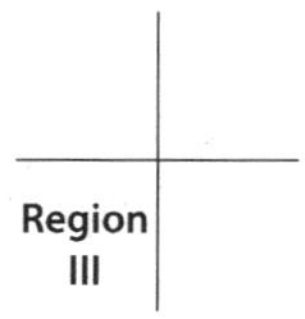

(1) The y-intercept of ℓ is 3

(2) The slope of ℓ is 4

To answer the question, you need:

(A) (1) alone
(B) (2) alone
(C) Both together
(D) Either one
(E) Can't answer

1

2

SOLUTION NEXT PAGE

DATA SUFFICIENCY	QUIZ	COORDINATE GEOMETRY

ALGEBRA

In the rectangular coordinate system shown, does the line ℓ (not shown) intersect Region III?

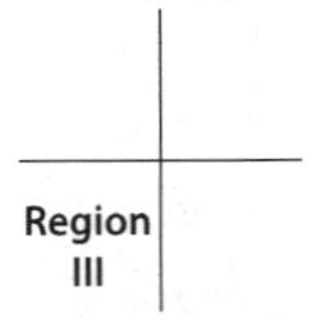

SOLUTION

For more on Coordinate Geometry, including the slope of the line, see Algebra, p. 144.

STEP 1 What is the question really asking or what do you need

If the slope is **positive**, the line crosses Region III.

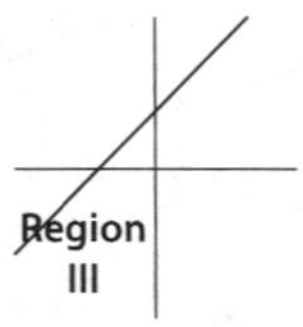

Positive slope

Line crosses R III

Positive slope

Line crosses R III

If the slope is **negative**, the line might cross Region III.

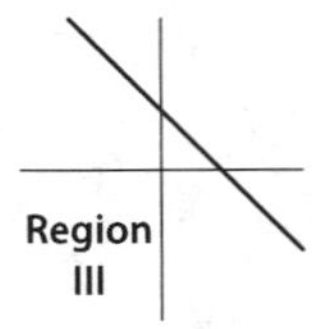

Negative slope

Doesn't cross R III

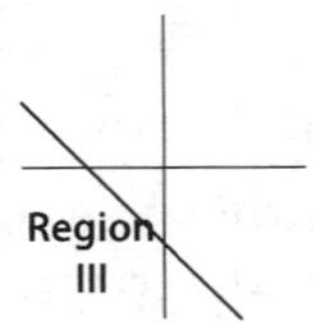

Negative slope

Line crosses R III

The question is really asking, "Does the line have a positive slope?"

STEP 2 Simplify one statement to determine what it tells you

(1) The y-intercept of ℓ is 3

From the drawings above, with a positive y-intercept, the line might or might not (the first and third drawings) cross Region III.

~~1~~ Statement (1) is not sufficient. Cross off the 1.

STEP 3 Simplify the other statement to determine what it tells you

(2) The slope of ℓ is 4

4 is positive, which answers the question.

(2) Statement (2) is sufficient. Circle the 2.

Answer is (B) 2 alone

QUIZ | PARABOLAS | DATA SUFFICIENCY

LIFELINE 63

To solve a parabola, find the x and y values of one point, and plug them into the equation.

ALGEBRA

In the rectangular coordinate system, if the constant k is a positive integer and $y = kx^2 - 2$, what is the value of k ?

(1) The graph intersects the x-axis at $(-\sqrt{2}, 0)$ and $(\sqrt{2}, 0)$

(2) When $x = 3, y = 7$

To answer the question, you need:

(A) (1) alone
(B) (2) alone
(C) Both together
(D) Either one
(E) Can't answer

1

2

Lifeline 63 is on p. 150

Remember
You know it's a parabola when the equation has some form of $y = x^2$.

For more on this, see Algebra p. 149.

SOLUTION NEXT PAGE

DATA SUFFICIENCY | QUIZ | PARABOLAS

ALGEBRA

In the rectangular coordinate system, if the constant k is a positive integer and $y = kx^2 - 2$, what is the value of k ?

SOLUTION

STEP 1 **What is the question really asking or what do you need**

If $y = kx^2 - 2$, then to solve for k, you need a single instance of x and y to plug into the equation.

The question really asks, "What are the (x, y) coordinates of one point on the graph?"

Remember
To solve a parabola, find the (x, y) coordinates of a single point.

For more on this, see Algebra, p.150.

STEP 2 **Simplify one statement to determine what it tells you**

(1) The graph intersects the x-axis at $(-\sqrt{2}, 0)$ and $(\sqrt{2}, 0)$

This gives the x and y coordinates of two points on the graph.

This answers the question, but to prove it:

Start with the equation	$y = kx^2 - 2$
Add the values of x and y	$0 = k(\sqrt{2})^2 - 2$
	$0 = k(2) - 2$
	$2 = k(2)$
	$k = 1$

Plugging $(-\sqrt{2})$ in for x gives the same answer.

1 Statement (1) is sufficient. Circle the 1.

STEP 3 **Simplify the other statement to determine what it tells you**

(2) When $x = 3$, $y = 7$

This statement also gives the x and y coordinates of a point on the graph.

Start with the equation	$y = kx^2 - 2$
Add the values of x and y	$7 = k(3)^2 - 2$
	$7 = k(9) - 2$
	$9 = k(9)$
	$k = 1$

2 Statement (2) is sufficient. Circle the 2.

Answer is (D) Either one

QUIZ	SIMULTANEOUS EQUATIONS	DATA SUFFICIENCY
LIFELINE 64	**You need two equations to solve for two unknowns.**	ALGEBRA

What is the value of $x + y$?

(1) $x + 2y = 4$

(2) $2x + y = 5$

To answer the question, you need:

(A) (1) alone
(B) (2) alone
(C) Both together
(D) Either one
(E) Can't answer

1

2

Lifeline 64 is on p. 153

SOLUTION NEXT PAGE

DATA SUFFICIENCY | QUIZ | SIMULTANEOUS EQUATIONS

ALGEBRA

What is the value of $x + y$?

SOLUTION

STEP 1 **What is the question really asking or what do you need**

To find the value of $x + y$, you need either
one equation that can be simplified to "$x + y =$ (a number)"
or two separate equations to solve for x and y.

STEP 2 **Simplify one statement to determine what it tells you**

(1) $x + 2y = 4$

This is a single equation that can't be simplified to "$x + y =$ (a number)"

~~1~~ Statement (1) is not sufficient. Cross off the 1.

STEP 3 **Simplify the other statement to determine what it tells you**

(2) $2x + y = 5$

This is also a single equation that can't be simplified to "$x + y =$ (a number)"

~~2~~ Statement (2) is not sufficient. Cross off the 2.

STEP 4 **Both 1 and 2 are crossed off, so check the statements together**

For more on simultaneous equations, see Algebra, p. 153.

To solve for $x + y$, combine the equations:

$$\begin{array}{r} x + 2y = 4 \\ +\ (2x + y = 5) \\ \hline 3x + 3y = 9 \\ \\ x + y = 3 \end{array}$$

Answer is (C) Both together

QUIZ | FACTORING EXPRESSIONS | DATA SUFFICIENCY

LIFELINE 66

With x^2, find the two possible values of x.

ALGEBRA

What is the value of x ?

(1) $(x + 2)(x - 3) = 0$

(2) $(x - 3)(x + 5) = 0$

To answer the question, you need:

(A) (1) alone
(B) (2) alone
(C) Both together
(D) Either one
(E) Can't answer

1

2

Lifeline 66 is on p. 161

SOLUTION NEXT PAGE

DATA SUFFICIENCY | QUIZ | FACTORING EXPRESSIONS

ALGEBRA

What is the value of x ?

SOLUTION

STEP 1 What is the question really asking or what do you need

What is the value of x ?

STEP 2 Simplify one statement to determine what it tells you

Remember
x has **one** value, and you don't know which it is.

For more on this, see Algebra, p. 61.

(1) $(x + 2)(x - 3) = 0$

x is either -2 or 3, but you don't know which one.

You need a single, numeric answer for the statement to be sufficient.

~~1~~

Statement (1) is not sufficient. Cross off the 1.

STEP 3 Simplify the other statement to determine what it tells you

(2) $(x - 3)(x + 5) = 0$

x is either 3 or -5, but you don't know which one.

~~2~~

Statement (2) is not sufficient. Cross off the 2.

STEP 4 Both 1 and 2 are crossed off, so check the statements together

If one statement says x is either -2 or 3,

and the other says x is either 3 or -5,

x must be equal to 3.

Answer is (C) Both together

QUIZ	MATH OPERATORS	DATA SUFFICIENCY
LIFELINE 69	**When working with new math operators, use real numbers for *x* and *y*.**	ALGEBRA

If the function Θ represents either addition, subtraction, multiplication, or division, which one does it represent?

(1) $a \Theta b = b \Theta a$ for all real numbers a and b

(2) $a \Theta 0 = a$ for all real numbers a

To answer the question, you need:

(A) (1) alone
(B) (2) alone
(C) Both together
(D) Either one
(E) Can't answer

1

2

Lifeline 69 is on p. 175

SOLUTION NEXT PAGE

DATA SUFFICIENCY | QUIZ | MATH OPERATORS

ALGEBRA

If the function Θ represents either addition, subtraction, multiplication, or division, which does it represent?

SOLUTION

STEP 1 **What is the question really asking or what do you need**

For example:

What is 6 Θ 3 ?	If addition	6 Θ 3 = 9
	If subtraction	6 Θ 3 = 3
	If multiplication	6 Θ 3 = 18
	If division	6 Θ 3 = 2

STEP 2 **Simplify one statement to determine what it tells you**

(1) $a \Theta b = b \Theta a$ for all real numbers *a* and *b*.

Does 6 Θ 3 = 3 Θ 6?	If addition	6 Θ 3 = 9	and	3 Θ 6 = 9
	If subtraction	6 Θ 3 = 3	but	3 Θ 6 = – 3
	If multiplication	6 Θ 3 = 18	and	3 Θ 6 = 18
	If division	6 Θ 3 = 2	but	3 Θ 6 = ½

Therefore, Θ represents either addition or multiplication.

~~1~~ Statement (1) is not sufficient, so cross off the 1.

STEP 3 **Simplify the other statement to determine what it tells you**

(2) $a \Theta 0 = a$ for all real numbers *a*.

Does 6 Θ 0 equal 6?	If addition	6 Θ 0 = 6	yes
	If subtraction	6 Θ 0 = 6	yes
	If multiplication	6 Θ 0 = 0	no
	If division	6 Θ 0 doesn't work, so no	

Therefore, Θ represents either addition or subtraction.

~~2~~ Statement (2) is not sufficient, so cross off the 2.

STEP 4 **Both 1 and 2 are crossed off, so check the statements together**

Statement (1) says Θ represents either addition or multiplication.

Statement (2) says it represents either addition or subtraction.

Therefore Θ must represent addition.

Answer is (C) Both together

QUIZ	TRIANGLES	DATA SUFFICIENCY
LIFELINE 80	**When you know two angles of a triangle, find the third angle.**	GEOMETRY

Lifeline 80 is on p. 217

In the figure shown, the measure of angle *BDA* is how many degrees greater than the measure of angle *BCD* ?

(1) The measure of angle *BCD* is 20°.

(2) Triangle *BCD* is isosceles.

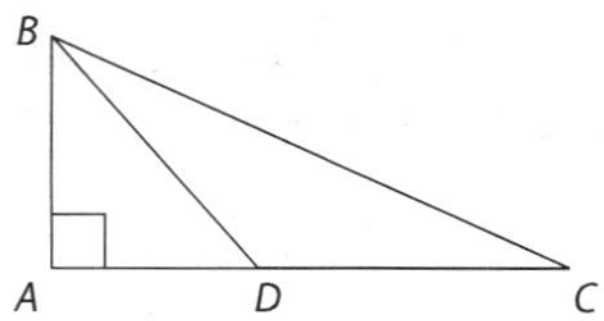

To answer the question, you need:

(A) (1) alone
(B) (2) alone
(C) Both together
(D) Either one
(E) Can't answer

1

2

SOLUTION NEXT PAGE

DATA SUFFICIENCY | **QUIZ** | **TRIANGLES / ANGLE TOTALS**

GEOMETRY

In the figure shown, the measure of angle *BDA* is how many degrees greater than the measure of angle *BCD* ?

SOLUTION

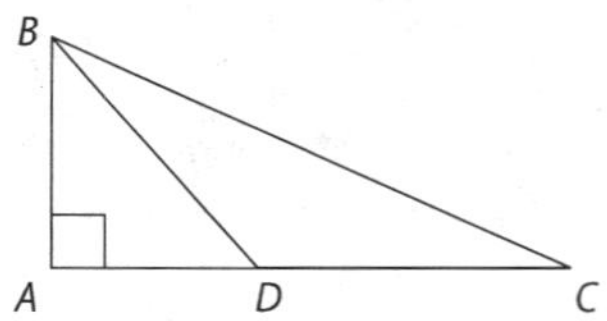

STEP 1 **What is the question really asking or what do you need**

To find this, you need to know the angles.
The question is really asking, "What are the measures of *BDA* and *BCD*?"

STEP 2 **Simplify one statement to determine what it tells you**

(1) The measure of angle *BCD* is 20° .

Statement (1) does not tell you the measure of angle *BDA*.

~~1~~ Statement (1) is not sufficient. Cross off the 1.

STEP 3 **Simplify the other statement to determine what it tells you**

(2) Triangle *BCD* is isosceles

This means that angles *BCD* and *CBD* have the same measure. Call this angle measure *x*

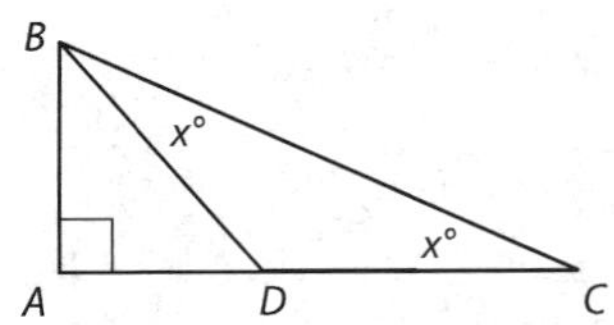

Because all triangles total 180°, angle BDC is (180 – 2x)°

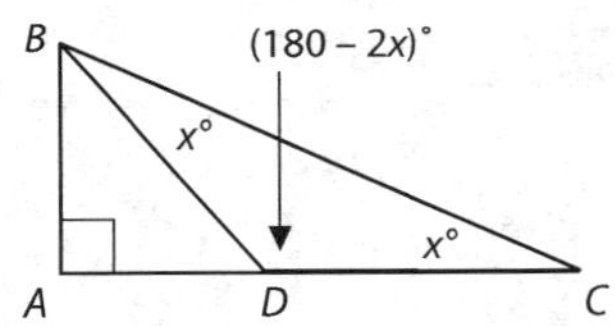

Because supplementary angles total 180°, angle *BDA* is 2x° .

Statement (2) gives you a ratio of the two angle measures in question. The measure of angle *BCD* is half that of angle *BDA*.

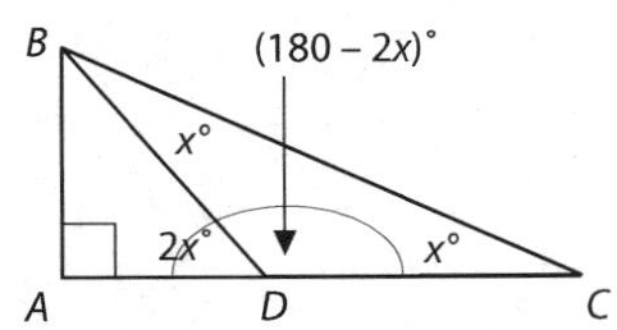

> **Remember**
> Supplementary angles total 180°.
>
> For more on this, see Geometry, p. 217.

However, without knowing the measure of one of the angles, you can't answer the question with this information alone.

~~2~~ Statement (2) is not sufficient. Cross off the 2.

STEP 4 **Both 1 and 2 are crossed off, so check the statements together**

Knowing that the measure of angle *BCD* is 20° , and the measure of angle *BDA* is twice that at 40° , you can answer the question.

Answer is (C) Both together

QUIZ | TRIANGLES / ISOSCELES | DATA SUFFICIENCY

LIFELINE 81

An isosceles triangle is symmetrical.

GEOMETRY

In isosceles triangle *PQR*, what is the measure of angle *P* ?

Lifeline 81 is on p. 221

(1) The measure of angle *Q* is 45°

(2) The measure of angle *R* is 90°

To answer the question, you need:

(A) (1) alone
(B) (2) alone
(C) Both together
(D) Either one
(E) Can't answer

1

2

SOLUTION NEXT PAGE

DATA SUFFICIENCY | QUIZ | TRIANGLES / ISOSCELES

GEOMETRY

In isosceles triangle *PQR*, what is the measure of angle *P* ?

SOLUTION

STEP 1 **What is the question really asking or what do you need**

In isosceles triangle *PQR*,
what is the measure of angle *P* ?

The question is asking,
"What are the measures of the angles,
and which angle is *P* ?"

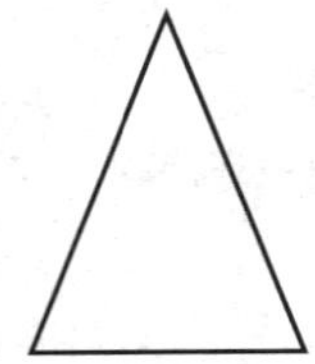

STEP 2 **Simplify one statement to determine what it tells you**

(1) The measure of angle *Q* is 45°

An isosceles triangle has two twin angles and one unique angle.

The 45° angle could be one of the **twin** angles, forming a 45°-45°-90° triangle.

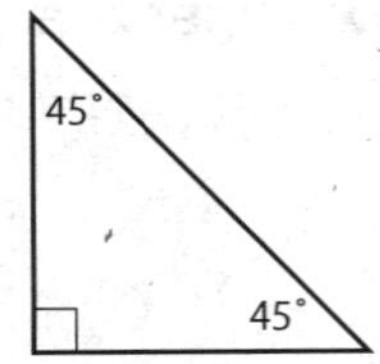

A 45° - 45° - 90° triangle

Or, the 45° angle could be the **unique** angle, making the other two angles 67.5°.

Without knowing which triangle is true or which angle *P* refers to, you can't answer the question.

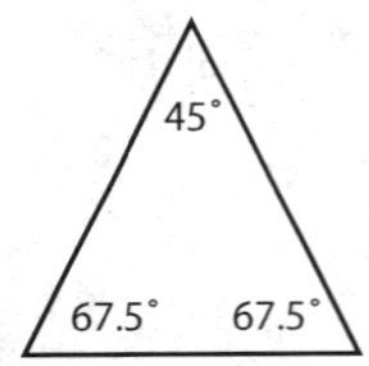

A 45° - 67.5° - 67.5° triangle

~~1~~ Statement 1 is not sufficient. Cross off the 1.

STEP 3 **Simplify the other statement to determine what it tells you**

(2) The measure of angle *R* is 90°

The 90° angle can't be one of the **twin** angles, because if the other angle were 90°, the shape wouldn't be a triangle.

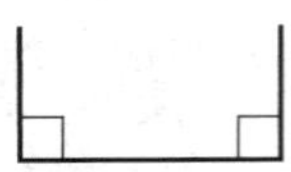
Not the case

This means the 90° angle, *R*, is the **unique** angle, and the remaining twin angles, *P* and *Q*, are each 45°.

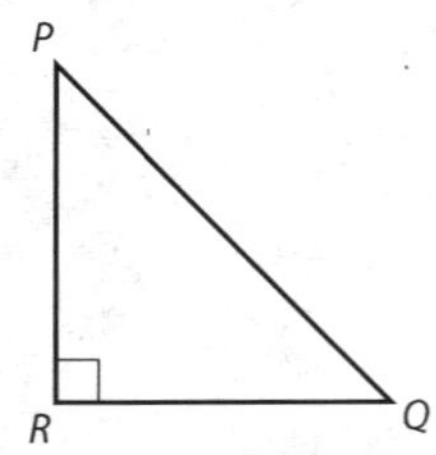

(2) Statement 2 is sufficient. Circle the 2.

Answer is (B) 2 alone

 thinktankSCHOLASTIC

QUIZ	TRIANGLES / 45-45-90	DATA SUFFICIENCY
LIFELINE 83	**With the 45-45-90 triangle, there are two side-length ratios, and one of them answers the question.**	GEOMETRY

What is the length of the hypotenuse of a certain right triangle?

(1) The side lengths adjacent to the right angle have a total length of 20.

(2) The triangle is an isosceles right triangle.

To answer the question, you need:

(A) (1) alone
(B) (2) alone
(C) Both together
(D) Either one
(E) Can't answer

1

2

Lifeline 83 is on p. 227

SOLUTION NEXT PAGE

DATA SUFFICIENCY | QUIZ | TRIANGLES / 45-45-90

GEOMETRY

What is the length of the hypotenuse of a certain right triangle?

SOLUTION

STEP 1 **What is the question really asking or what do you need**

Drawing a picture will help organize your thoughts, even if you don't have much information.

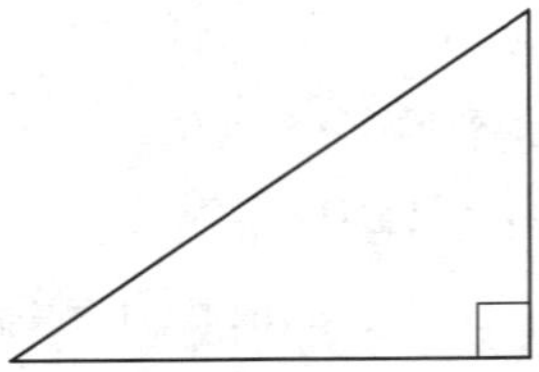

STEP 2 **Simplify one statement to determine what it tells you**

(1) The side lengths adjacent to the right angle have a total length of 20.

The side lengths could still be anything (8 + 12, 9 + 11, etc.), so you can't find the hypotenuse, with that information.

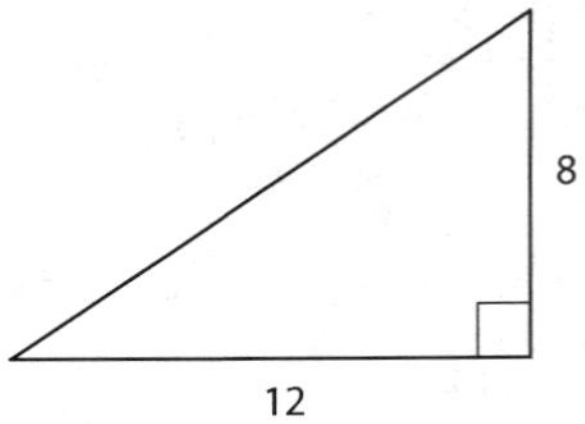

~~1~~ Statement (1) is not sufficient. Cross off the 1.

STEP 3 **Simplify the other statement to determine what it tells you**

(2) The triangle is an isosceles right triangle.

This statement provides the side-length ratio, but no side lengths for scale.

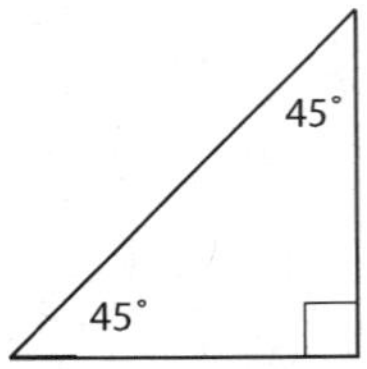

~~2~~ Statement (2) is not sufficient. Cross off the 2.

STEP 4 **Both 1 and 2 are crossed off, so check the statements together**

An isosceles right triangle with side lengths of 20 has a hypotenuse of $10\sqrt{2}$

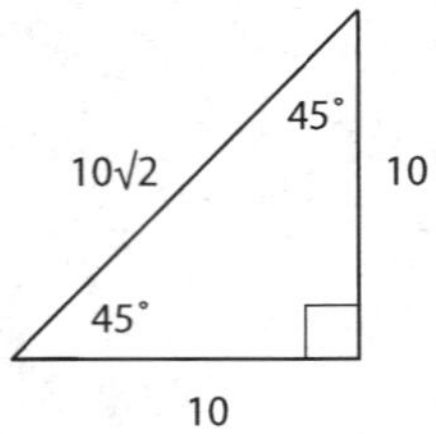

Remember
When working with the 45-45-90 triangle, use the side ratios.

For more on this, see Geometry, p. 227.

Answer is (C) Both together

QUIZ	PARTS OF CIRCLES	DATA SUFFICIENCY
LIFELINE 87	**To find the arc or the sector, measure the whole circle and take the fraction.**	GEOMETRY

In the circle shown, if the radius is 9, what is the length of minor arc *CD* ?

Lifeline 87 is on p. 243

(1) Angle $ADB = 20°$

(2) Segments *BC* and *AD* are parallel

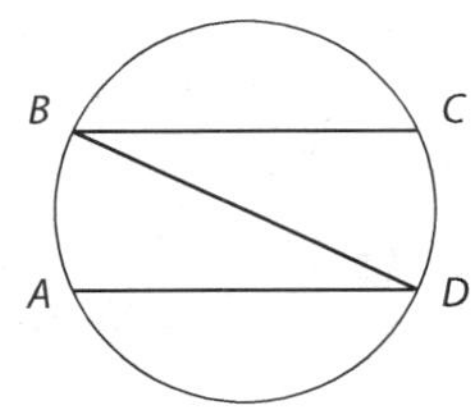

To answer the question, you need:

(A) (1) alone
(B) (2) alone
(C) Both together
(D) Either one
(E) Can't answer

1

2

SOLUTION NEXT PAGE

DATA SUFFICIENCY	QUIZ	PARTS OF CIRCLES

GEOMETRY

In the circle shown, if the radius is 9, what is the length of minor arc CD ?

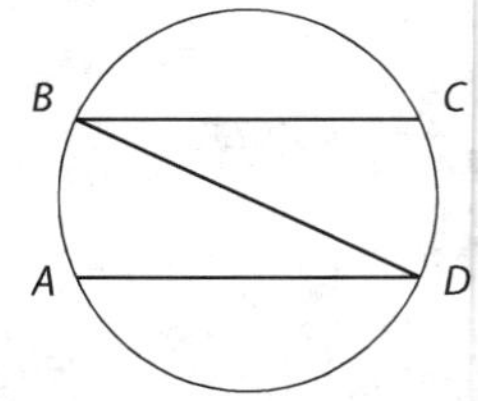

SOLUTION

STEP 1 **What is the question really asking or what do you need**

To find the length of an arc, you need the circumference of the circle and the degree measure of the arc. You know that the circumference is 18π because the radius is 9.

> **Remember**
> When the angle is on the circle, the arc has twice the degree measure.
>
> For more on this, see Geometry, p. 247

The question really asks, "What is the degree measure of arc CD ?"

STEP 2 **Simplify one statement to determine what it tells you**

(1) Angle $ADB = 20°$

This doesn't give you enough information to find the degree measure of arc CD .

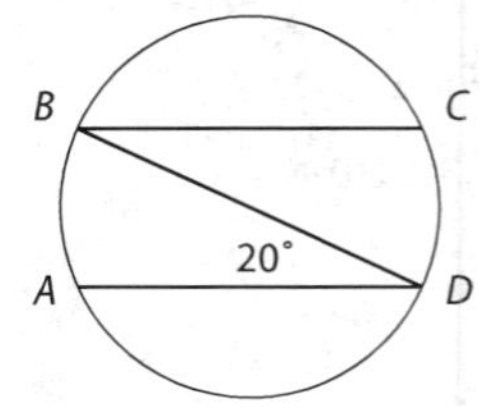

~~1~~

Statement (1) is not sufficient. Cross off the 1.

STEP 3 **Simplify the other statement to determine what it tells you**

2) Segments BC and AD are parallel

When a third line (segment BD) crosses crosses two parallel lines, it creates sets of equal angles. This means that angles BDA and CBD are equal.

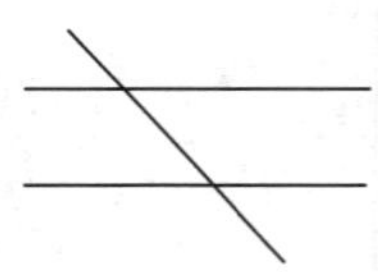

However, without Statement (1), you don't know the angles.

~~2~~

Statement (2) is not sufficient. Cross off the 2.

STEP 4 **Both 1 and 2 are crossed off, so check the statements together**

Knowing the segments are parallel, and given that angle ADB is 20°, you know that angle CBD is also 20° .

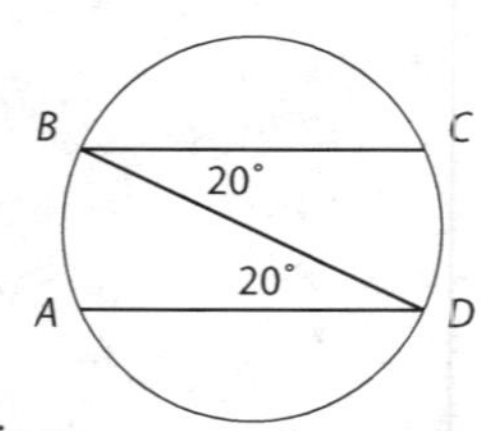

Arc BD therefore is 40°, which answers the question.

Answer is (C) Both together

QUIZ	INDIVIDUALS—SAME RATE	DATA SUFFICIENCY
LIFELINE 96	**With individuals working at the same rate, find the fraction of the job one individual does in one hour or day.**	WORD PROBLEMS

How many days would it take 6 machines, each working at the same rate, to complete a certain job?

(1) 4 of these machines, each working at the same rate, can complete the job in 12 days.

(2) 8 of these machines, each working at the same rate, can complete the job in half the amount of time that 4 of these machines, each working at the same rate, can complete the job.

To answer the question, you need:

(A) (1) alone
(B) (2) alone
(C) Both together
(D) Either one
(E) Can't answer

1

2

Lifeline 96 is on p. 273

SOLUTION NEXT PAGE

DATA SUFFICIENCY | QUIZ | INDIVIDUALS—SAME RATE

WORD PROBLEMS

How many days would it take 6 machines, each working at the same rate, to complete a certain job?

SOLUTION

STEP 1 **What is the question really asking or what do you need**

The question really asks, "How much of the job does each machine, by itself, complete in one day?"

For more on individuals working at the same rate, see Word Problems, p. 273.

STEP 2 **Simplify one statement to determine what it tells you**

(1) 4 of these machines, each working at the same rate, can complete the job in 12 days.

Finding the work that each machine does in one day

Step	Demo
1. Find the fraction that the team does in one day	$\frac{1}{12}$
2. Find the fraction of that day's work that each of the 4 machines does in one day	$\frac{1}{4}$
3. Multiply the fractions	$\frac{1}{12} \times \frac{1}{4} = \frac{1}{48}$

Knowing one machine completes $\frac{1}{48}$ of the job in one day,

6 machines would need 8 days to complete the job.

1 Statement (1) is sufficient. Circle the 1.

STEP 3 **Simplify the other statement to determine what it tells you**

(2) 8 of these machines, each working at the same constant rate, can complete the job in half the amount of time that 4 of these machines, each working at the same constant rate, can complete the job.

But you still don't know the fraction of the job each machine completes in one day.

~~2~~ Statement (2) is not sufficient. Cross off the 2.

Answer is (A) 1 alone

QUIZ	CONVERSIONS	DATA SUFFICIENCY
LIFELINE 98	**When working a conversion question, set up the fractions so that you can cancel everything out.**	WORD PROBLEMS

Did it take a toy car less than 2 minutes to roll 3 meters?
(1 meter = 100 centimeters.)

(1) On average, the toy car rolled less than 3 centimeters per second.

(2) On average, the toy car rolled more than 2 centimeters per second.

To answer the question, you need:

(A) (1) alone
(B) (2) alone
(C) Both together
(D) Either one
(E) Can't answer

Lifeline 98 is on p. 283

SOLUTION NEXT PAGE

DATA SUFFICIENCY | **QUIZ** | **CONVERSIONS**

WORD PROBLEMS

SOLUTION

STEP 1 What is the question really asking or what do you need

To roll 3 meters, did it take the toy car less than 2 minutes?

On average, was the toy car's speed greater than $\frac{\text{3 meters}}{\text{2 minutes}}$?

Convert meters per minute to centimeters per second
to match Statements (1) and (2)

Steps	**Example**
1. Set up the fractions	$\left[\frac{\text{3 meters}}{\text{2 minutes}}\right]\left[\frac{\text{100 cm}}{\text{1 meter}}\right]\left[\frac{\text{1 minute}}{\text{60 seconds}}\right]$
2. Cancel the fractions a. Units first	$\left[\frac{\text{3 \cancel{meters}}}{\text{2 \cancel{minutes}}}\right]\left[\frac{\text{100 cm}}{\text{1 \cancel{meter}}}\right]\left[\frac{\text{1 \cancel{minute}}}{\text{60 seconds}}\right]$
b. Numbers second	$\left[\frac{\cancel{3}}{\cancel{2}}\right]\left[\frac{\cancel{100}\text{ cm}}{1}\right]\left[\frac{1}{\cancel{60}\text{ seconds}}\right]$ (100 = 10 × 10 = (5 × 2) × 10; 60 = 6 × 10 = (2 × 3) × 10)
3. Multiply the fractions	$\left[\frac{1}{1}\right]\left[\frac{\text{5 cm}}{1}\right]\left[\frac{1}{\text{2 seconds}}\right] = \left[\frac{\text{5 cm}}{\text{2 seconds}}\right] = \left[\frac{\text{2.5 cm}}{\text{1 seconds}}\right]$

The question is really asking:

"On average, did the toy car roll more than 2.5 cm per sec?"

STEP 2 Simplify one statement to determine what it tells you

(1) On average, the toy car rolled less than 3 cm per sec

The toy car could have rolled 2.1 cm per sec or 2.8 cm per sec

~~1~~ Statement (1) is not sufficient. Cross off the 1.

STEP 3 Simplify the other statement to determine what it tells you

(2) On average, the toy car rolled more than 2 cm per sec

Again, the car could have rolled 2.1 cm per sec or 2.8 cm per sec

~~2~~ Statement (2) is not sufficient. Cross off the 2.

STEP 4 Both 1 and 2 are crossed off, so check the statements together

The car still could have rolled either 2.1 cm per sec or 2.8 cm per sec,
so the question can't be answered.

Answer is (E) Can't answer

QUIZ	SET DISTRIBUTION	DATA SUFFICIENCY
LIFELINE 99	**When working on set distribution, use the Set formula.**	WORD PROBLEMS

Lifeline 99 is on p. 287

If each student at a school is enrolled in either Math, German, or both, how many students are at the school?

(1) The number of students enrolled in both Math and German is 10.

(2) The number of students enrolled in Math minus the number of students enrolled in German is 15.

To answer the question, you need:

(A) (1) alone
(B) (2) alone
(C) Both together
(D) Either one
(E) Can't answer

1

2

SOLUTION NEXT PAGE

DATA SUFFICIENCY | QUIZ | SET DISTRIBUTION

WORD PROBLEMS

If each student at a school is enrolled in either Math, German, or both, how many students are at the school?

SOLUTION

STEP 1 | **What is the question really asking or what do you need**

Use the Set formula:

$$\text{total} = (\text{set}_1 + \text{set}_2) - \text{overlap}(_{1\ \&\ 2})$$

The question is really asking, "How many students are in Math, German, and both classes?"

STEP 2 | **Simplify one statement to determine what it tells you**

(1) The number of students enrolled in both Math and German is 10.

This tells you the overlap, but not the sets.

Statement (1) is not sufficient. Cross off the 1.

~~1~~

STEP 3 | **Simplify the other statement to determine what it tells you**

(2) The number of students enrolled in Math minus the number of students enrolled in German is 15.

If the statement gave you Math *plus* German, you could use the formula, but knowing math *minus* German doesn't help.

Math and German could have 30 and 15 students, respectively, or 85 and 70 students.

Statement (2) is not sufficient. Cross off the 2.

~~2~~

STEP 4 | **Both 1 and 2 are crossed off, so check the statements together**

You can fill in more of the formula, but Math and German could still have 30 and 15 students or 85 and 70 students.

Taken together, you can't answer this question.

Answer is (E) Can't answer

QUIZ	SETS IN TABLES	DATA SUFFICIENCY
LIFELINE 100	**When working a set in table question, set up the table to do your work.**	WORD PROBLEMS

Of the applicants for a job, what percent have at least 6 years' experience and a degree?

(1) 42% of the applicants have at least 6 years' experience.

(2) 38% of the applicants have a degree.

To answer the question, you need:

(A) (1) alone
(B) (2) alone
(C) Both together
(D) Either one
(E) Can't answer

1

2

Lifeline 100 is on p. 291

SOLUTION NEXT PAGE

DATA SUFFICIENCY | QUIZ | SETS IN TABLES

WORD PROBLEMS

SOLUTION

Of the applicants for a job, what percent have at least 6 years' experience and a degree?

STEP 1 What is the question really asking or what do you need

This is a Set-in-Table question, so draw the table.

1. Set up the table with the groups on the outside, leaving room to add the numbers.

	Experience Yes / No
Degree	
Yes	____ ?
No	

To complete the table, you need the numbers on the top and side and a single number inside.

STEP 2 Simplify one statement to determine what it tells you

(1) 42% of the applicants have at least 6 years' experience.

This tells you the numbers on the top, (because 58% don't have experience) but nothing else.

Statement (1) is not sufficient. Cross off the 1.

~~1~~

STEP 3 Simplify the other statement to determine what it tells you

(2) 38% of the applicants have a degree.

This tells you the numbers on the side, (because 62% don't have degrees) but nothing else.

Statement (2) is not sufficient. Cross off the 2.

~~2~~

STEP 4 Both 1 and 2 are crossed off, so check the statements together

2. Taken together, you can add to the table:

	Experience Yes **42%** / No **58%**
Degree	
Yes **38%**	____ ?
No **62%**	

For more on sets in tables, see Word Problems, p. 291.

You need a single number to go inside, but without that, you can't answer the question.

Answer is (E) Can't answer

QUIZ | UNORDERED COUNTING | DATA SUFFICIENCY

LIFELINE 104

When the order doesn't matter, use the formula.

WORD PROBLEMS

A certain number of students are to be selected from a class to work on a project. How many different selections of students are possible?

(1) 3 students are to be selected for the project.

(2) Half of the students in the class will be selected for the project.

To answer the question, you need:

(A) (1) alone
(B) (2) alone
(C) Both together
(D) Either one
(E) Can't answer

1

2

Lifeline 104 is on p. 316

SOLUTION NEXT PAGE

DATA SUFFICIENCY	QUIZ	UNORDERED COUNTING

WORD PROBLEMS

A certain number of students are to be selected from a class to work on a project. How many different selections of students are possible?

SOLUTION

For more on this, see Word Problems, p. 316

STEP 1 What is the question really asking or what do you need

With this Unordered Counting question, use the formula:

$$\text{\# of possibilities} = \frac{\text{(whole group)!}}{\text{(those taken)! (those left over)!}}$$

Because (those taken) + (those left over) = (whole group) you need two of these three numbers.

STEP 2 Simplify one statement to determine what it tells you

(1) 3 students are to be selected for the project.

3 students are taken, but you still don't know the number left over or the whole group.

~~1~~ Statement (1) is not sufficient. Cross off the 1.

STEP 3 Simplify the other statement to determine what it tells you

(2) Half of the students in the class will be selected for the project.

Half the students are taken and half are left over, but this doesn't give you any numbers.

~~2~~ Statement (2) is not sufficient. Cross off the 2.

STEP 4 Both 1 and 2 are crossed off, so check the statements together

If half the students are taken, and that half is 3, then the number of students left over also is 3, making the class size (whole group) 6.

Now you can complete the equation: # of possibilities $= \frac{6!}{3!3!}$

Answer is (C) Both together

You don't have to actually solve it, but here's the solution:

$$\text{\# of possibilities} = \frac{6!}{3!3!} = \frac{6 \times 5 \times 4 \times \cancel{3!}}{3! \times \cancel{3!}} = \frac{\cancel{6} \times 5 \times 4}{\cancel{3 \times 2} \times 1} = 5 \times 4 = 20$$

thinktankSCHOLASTIC

QUIZ | WEIGHTED AVERAGES | DATA SUFFICIENCY

LIFELINE 106

With the weighted average, you can find the ratio.

WORD PROBLEMS

At a certain company, the average (arithmetic mean) number of accrued vacation days is 10 days for the male employees and 15 days for the female employees. What is the number of male employees?

(1) There are 20 female employees at the company.

(2) The average number of accrued vacation days for all of the company's employees is 12 days.

To answer the question, you need:

(A) (1) alone
(B) (2) alone
(C) Both together
(D) Either one
(E) Can't answer

Lifeline 106 is on p. 325

1

2

SOLUTION NEXT PAGE

DATA SUFFICIENCY | **QUIZ** | **WEIGHTED AVERAGES**

WORD PROBLEMS

At a certain company, the average (arithmetic mean) number of accrued vacation days is 10 days for the male employees and 15 days for the female employees. What is the number of male employees?

SOLUTION

STEP 1 What is the question really asking or what do you need

This is a weighted average question, so you need the overall average to find the ratio. Once you have the ratio, you need one number to find the other number.

For more on completing the ratio, see Arithmetic, p. 71.

STEP 2 Simplify one statement to determine what it tells you

(1) There are 20 female employees at the company.
This helps, but you still don't have the average.

~~1~~ Statement (1) is not sufficient. Cross off the 1.

STEP 3 Simplify the other statement to determine what it tells you

(2) The average number of accrued vacation days for the company's male and female employees is 12 days. With this overall average, you can find the ratio:

1. Find the distance of each group from the average

Males	Average	Females
10	12	15

← 2 → ← 3 →

2. Cross multiply

Males	Average	Females
10	12	15

2 3

3. Set up the fraction

$$\frac{\text{Males}}{\text{Females}} = \frac{3}{2}$$

~~2~~ Without a number to place in the ratio, you can't answer the question. Cross off the 2.

STEP 4 Both 1 and 2 are crossed off, so check the statements together

With the ratio and an amount, you can answer the question:

$$\frac{\text{Males}}{\text{Females}} = \frac{3}{2} = \frac{30}{20}$$

Answer is (C) Both together

HOMEWORK QUESTIONS

13th Editon *GMAT Review Book,* pp. 152 -185, question numbers

GMAT REVIEW

Arithmetic

Odd times Even
7 _____
14 _____
27 _____
32 _____
72 _____
111 _____
173 _____

Divisibility
58 _____
71 _____
80 _____
83 _____

Units Digit
75 _____
170 _____

Consecutive Integers
18 _____
64 _____
95 _____

Fractions
4 _____
10 _____
50 _____
82 _____
92 _____
131 _____
156 _____
171 _____
174 _____

Decimals
31 _____
63 _____
133 _____
148 _____
157 _____
169 _____

Percents
8 _____
28 _____
37 _____
40 _____
51 _____
55 _____
61 _____
62 _____
76 _____
77 _____
91 _____
127 _____
141 _____
150 _____
151 _____
161 _____

Ratios
23 _____
26 _____
47 _____
65 _____
81 _____
88 _____
114 _____
116 _____
135 _____
158 _____
168 _____

Exponents
41 _____
53 _____
67 _____
97 _____
160 _____
162 _____

Radicals
15 _____
172 _____

Averages and Medians
20 _____
84 _____
87 _____
110 _____
120 _____
121 _____
139 _____
144 _____
146 _____
154 _____
163 _____

Range
123 _____

Standard Deviation
38 _____
109 _____

Algebra

Absolute Value
1 _____
159 _____

Number Line
6 _____
69 _____
112 _____

Inequalities
13 _____
46 _____
48 _____
52 _____
85 _____
90 _____
101 _____
105 _____
164 _____
167 _____

Coordinate Geometry
11 _____
74 _____
129 _____

Simultaneous Equations
9 _____
17 _____
36 _____
44 _____
45 _____
54 _____
57 _____
59 _____
60 _____
78 _____
98 _____
103 _____
124 _____
132 _____
142 _____
153 _____

Factoring Expressions
86 _____
99 _____

Math Operators
96 _____
104 _____
118 _____
136 _____

Sequences
24 _____

Geometry

Triangles
19 _____
56 _____
73 _____
79 _____
113 _____
119 _____
128 _____
149 _____
152 _____
155 _____
165 _____

Quadrilaterals
42 _____
145 _____
166 _____

Circles
100 _____
117 _____
122 _____

3-D Shapes
30 _____
35 _____
102 _____
130 _____
68 _____
115 _____

Word Problems

Work Process
12 _____
68 _____
115 _____

Rate / Distance
22 _____
106 _____
108 _____

Conversions
16 _____
107 _____
140 _____

Counting Methods
89 _____

Set Distribution
5 _____
49 _____
93 _____

Sets in Tables
2 _____
21 _____
29 _____
34 _____
43 _____
66 _____
94 _____
134 _____
137 _____
138 _____

Probability
3 _____
39 _____
125 _____

Weighted Averages
33 _____
70 _____

Setting up Equations
25 _____
126 _____
143 _____
147 _____

DATA SUFFICIENCY	HOMEWORK ANSWERS
GMAT REVIEW	**13th Editon** *GMAT Review Book,* pp. 152 -185, question numbers

Arithmetic

Odd times Even
7 ______ D
14 ______ B
27 ______ D
32 ______ D
72 ______ B
111 ____ D
173 ____ C

Divisibility
58 ______ A
71 ______ D
80 ______ A
83 ______ E

Units Digit
75 ______ D
170 ____ A

Consecutive Integers
18 _____ D
64 ______ D
95 ______ D

Fractions
4 _______ A
10 ______ A
50 ______ B
82 ______ D
92 ______ B
131 ____ B
156 ____ A
171 ____ A
174 ____ B

Decimals
31 ______ B
63 ______ B
133 ____ A
148 ____ B
157 ____ D
169 ____ D

Percents
8 _______ A
28 ______ D
37 ______ A
40 ______ A
51 ______ B
55 ______ E
61 ______ C
62 ______ C
76 ______ D
77 ______ C
91 ______ B
127 _____ B
141 _____ B
150 _____ E
151 _____ D
161 _____ D

Ratios
23 ______ D
26 ______ B
47 ______ C
65 ______ C
81 ______ E
88 ______ A
114 _____ E
116 _____ D
135 _____ B
158 _____ B
168 _____ B

Exponents
41 ______ C
53 ______ A
67 ______ B
97 ______ C
160 _____ A
162 _____ B

Radicals
15 ______ D
172 _____ B

Averages and Medians
20 ______ B
84 ______ C
87 ______ D
110 _____ B
120 _____ C
121 _____ A
139 _____ C
144 _____ B
146 _____ D
154 _____ B
163 _____ D

Range
123 _____ E

Standard Deviation
38 ______ B
109 _____ C

Algebra

Absolute Value
1 _______ B
159 _____ D

Number Line
6 _______ E
69 ______ A
112 _____ D

Inequalities
13 ______ A
46 ______ E
48 ______ B
52 ______ D
85 ______ D
90 ______ E
101 _____ A
105 _____ B
164 _____ A
167 _____ A

Coordinate Geometry
11 ______ C
74 ______ A
129 _____ E

Simultaneous Equations
9 C
17 ______ B
36 ______ D
44 ______ D
45 ______ B
54 ______ C
57 ______ E
59 ______ D
60 ______ D
78 ______ A
98 ______ E
103 _____ D
124 _____ E
132 _____ A
142 _____ D
153 _____ D

Factoring Expressions
86 ______ C
99 ______ B

Math Operators
96 ______ C
104 _____ D
118 _____ D
136 _____ D

Sequences
24 ______ A

Geometry

Triangles
19 ______ A
56 ______ A
73 ______ A
79 ______ C
113 _____ A
119 _____ D
128 _____ D
149 _____ E
152 _____ D
155 _____ A
165 _____ D

Quadrilaterals
42 ______ E
145 _____ D
166 _____ C

Circles
100 _____ A
117 _____ D
122 _____ D

3-D Shapes
30 ______ E
35 ______ D
102 _____ E
130 _____ C
68 ______ A
115 _____ B

Word Problems

Work Process
12 ______ B
68 ______ A
115 _____ B

Rate / Distance
22 ______ D
106 _____ D
108 _____ E

Conversions
16 ______ C
107 _____ D
140 _____ C

Counting Methods
89 ______ A

Set Distribution
5 _______ E
49 ______ C
93 ______ C

Sets in Tables
2 _______ E
21 ______ E
29 ______ C
34 ______ E
43 ______ C
66 ______ D
94 ______ A
134 _____ A
137 _____ E
138 _____ D

Probability
3 _______ A
39 ______ E
125 _____ B

Weighted Averages
33 ______ B
70 ______ C

Setting up Equations
25 ______ E
126 _____ E
143 _____ C
147 _____ E

Tear Sheets

Five things to do so you don't get Slammed on test day.

You are about to jump into many hours of study. However, on race day, it will be these top five that make all the difference. Author's comment: "I've been teaching this test for five years. These are not speculation. They are observation."

1. Build your stamina

At four hours, the GMAT challenges your stamina as much as anything else. If you can't maintain test-level concentration for the entire exam, you'll start making all kinds of mistakes and getting the answers wrong.

2. Practice the essay

The test begins with an essay. Sure you can handle the math and verbal parts, but can you do so after struggling with an essay? There's no writer's block like there is in the testing center, and you need to get through this without burning too much energy, so that you have the strength to take on the rest of the test. This is an important part of building your stamina.

3. Use actual GMAT practice tests

The actual GMAT practice test is exactly like the real thing. The layout, the fonts, the colors: all true to form. The software covers the correct scope of material: For example, you won't be asked the volume of a cone, because the real GMAT doesn't ask that. Also, the actual GMAT practice test uses the same computer-based scoring system as the real test, so that you get a true sense of your abilities. The software is a free download from mba.com.

THINGS TO CONSIDER

4. Practice under pressure

Working test questions in class or at home is not the same as working them on test day, with high stakes and a timer. You can't simulate the testing center, but you can use a stop-watch to create pressure and fellow test-takers to bring competition—then you'll see what types of mistakes you make while competing and rushing, and you'll fix them.

5. Finish the test

The single most devastating way to kill your score is to let the clock run out before reaching the last question. The GMAT significantly lowers your score if you don't finish the test, so when there are three minutes left, stop what you're doing and click to the end.

9 Core Strategy LifeLines

1 If you're not sure when to take your GMAT, check your timeline, and schedule it now.

2 If you become exhausted, which is normal, take a two-week break from studying.

3 When you think you're ready for the GMAT. Take a practice test.

4 When there are three minutes left, stop working and click to the end.

5 If you're stuck on a question, take a guess, move on, and recover later in the section.

6 Forget the time per question. Focus on the first half of the section, and click through the second half.

7 If you find yourself quickly getting exhausted, build your stamina by studying in 3-hour stretches.

8 The morning of the exam, warm up with simple math and reading, so you're not starting with a GMAT question.

9 Control your test-day anxiety by making the GMAT testing-center experience as familiar as possible.

46 Arithmetic LifeLines

10 When x is a multiple of a number, use the three lowest multiples of the number.

11 When x is odd or even, use real odd or even numbers, such as 2 or 3.

12 When x divided by a number has a certain remainder, find some possible values of x.

13 When reducing a fraction, don't use long division: instead, factor and cancel.

14 When factoring a number with zeros, such as 300, always start with the zeros: 3×100

15 When dividing a number, use the number properties instead of long division.

16 When finding the "units" digit of the answer, use the "units" digits of the numbers you multiplied.

17 When working with consecutive integers, find the average.

18 When finding all the consecutive integers, start with a dash for each integer.

19 To find the sum of consecutive integers, multiply the average by the number of terms.

20 To quickly count the consecutive integers, start by subtracting the lowest from the highest.

21 When working with fractions, even just adding or subtracting, give the fractions a common denominator.

22 When working with fractions, cancel what you can.

23 When dividing fractions, flip the second fraction and multiply.

24 When two fractions equal each other, and one has an x, cross-multiply.

25 When there's a 1 on top of a fraction, flip the fraction.

26 When adding or subtracting numbers with decimals, line the decimal points up.

27 When multiplying numbers with decimals, count the decimal spaces.

28 When multiplying one number with a decimal and one with zeros, move the decimals.

113 LIFELINE LIST

29 When dividing numbers with decimals, move the decimals of both numbers to the right.

30 With scientific notation, move the decimal the number of spaces shown by the exponent.

31 When working with a percent, convert it to a decimal or fraction to make it easier.

32 To find the a percent of a percent, convert the percents to decimals or fractions, and multiply.

33 To find the percent of change, divide the amount of change by the original amount.

34 When the GMAT asks for the percent of a group, set it up as a fraction equal to $\frac{x}{100}$

35 To complete a ratio, set it up as a fraction with x as the missing number.

36 When finding numbers in a ratio, set the ratio up as an equation.

37 When finding the new ratio, set up the old ratios as fractions equal to each other.

38 When combining ratios, stack the part that they have in common.

39 When a number has a negative exponent, place a 1 on top.

40 When multiplying numbers with exponents, add the exponents.

41 When dividing numbers with exponents, subtract the exponents.

42 When taking the exponent of an exponent, multiply the exponents.

43 When multiplying or dividing numbers with different exponents, add the exponents.

44 When multiplying or dividing numbers with the same exponent, keep the exponent.

45 When simplifying a square root, factor it out.

46 When simplifying a cube root, look for triple factors.

47 When multiplying or dividing numbers with radicals, ignore the radicals.

48 When estimating a root, find the whole numbers just above and below the root.

49 When completing a radical, factor it out.

50 When working with the average, set up the equation with what you have.

51 When working with the median, put the numbers in order.

52 To find the mode, put the numbers in order.

53 To find the range, subtract the lowest number from the highest number.

54 When working with a standard deviation, start with the mean.

55 When comparing standard deviations, use the average deviation.

20 Algebra LifeLines

56 When working with the number line, draw the number line.

57 When x is inside the absolute value, place it on both sides of zero on the number line.

58 When working with an inequality, for the most part treat it like an equal sign.

59 When drawing the line from an equation, pick a value for x, and find the corresponding value of y.

60 To find the equation of a line, find the y-intercept and the slope.

61 When finding the slope of a line, use "rise over run."

62 When converting a linear equation to the form "$y = mx + b$" , basically set y equal to everything else.

63 To solve a parabola, find the x and y values of one point,and plug them into the equation.

64 You need two equations to solve for two unknowns.

65 When multiplying expressions, multiply everything in one bracket by everything in the other bracket.

66 With x^2, find the two possible values of x.

67 When factoring terms, factor out the largest item that's in each term.

68 When factoring exponents, start with the smaller exponent.

69 When working with new math operators, use real numbers for x and y.

70 When working a sequence, find the pattern.

71 When working with unknowns, use real numbers to set up the equation, but stick with letters to do the work.

72 When solving $f(x)$ functions, treat the $f(x)$ like a y.

73 When finding the $f(x)$ function, use real numbers for x.

74 When trying all the answer choices, work the first one, then the last one, then work your way up.

75 When up against complex math, use simple math to find the pattern.

19 Geometry LifeLines

76 With the measure of one angle, find the measure of the other three angles.

77 To find the angle totals, start by counting the angles.

78 To find the possible third side length of a triangle, draw a picture or use the formula.

79 For the area of a triangle, find the base and the height.

80 When you know two angles of a triangle, find the third angle.

81 When the angles are given as letters, use those letters to represent the unknown angle.

82 When the base is half the hypotenuse, think of the mnemonic 1- 2 - 3

83 With the 45-45-90 triangle, there are two side-length ratios, and one of them answers the question.

84 To measure the area of most quadrilaterals, find the base and the height.

85 When working with circles, leave *pi* as the symbol π, not 3.14

86 To find the surrounding area of two shapes, subtract the smaller shape from the larger shape.

87 To find the arc or the sector, measure the whole circle and take the fraction.

88 When the angle is on the circle, the arc has twice the degree measure.

89 When the shapes have the same perimeter, set the perimeters equal to each other.

90 When the shapes have the same area, set the areas equal to each other.

113 LIFELINE LIST

91 When working with the rectangular solid, remember that all the sides are rectangles.

92 When working with the cube, use the *edge* formulas.

93 When working with the cylinder, the volume is the circle's area times the cylinder's height.

94 When working with a different 3-D shape, draw a picture to see how it fits with the basic shapes.

13 Word Problems LifeLines

95 With individuals working at different rates, find the fraction of the job each individual does in one hour or day.

96 With individuals working at the same rate, find the fraction of the job one individual does in one hour or day.

97 To find the average speed, place the entire distance over the entire time (in hours).

98 When working a conversion question, set up the fractions so that you can cancel everything out.

99 When working on set distribution, use the Set formula.

100 When working a set in table question, set up the table to do the work.

101 When finding the number of possible outcomes, start by placing a dash for each event.

102 When working with factorials, factor them out and cancel.

103 When the order matters, place a dash for each event, and write the number of possible outcomes for that event.

104 When the order doesn't matter, use the formula.

105 When finding probability, find the number of possible outcomes.

106 With the weighted average, you can find the ratio.

107 When an equation has two unknowns, and finding one tells you the other, use x and (total – x).

6 Data Sufficiency LifeLines

108 When answering a data sufficiency question use these answer choices.

109 When checking the statements pay attention to how they fit together.

110 When the statements repeat each other, the only possible answers are (D) Either one or (E) Can't answer.

111 When working a Data Sufficiency question, cover up the statements and simplify the question.

112 When evaluating a complicated statement, simplify it first.

113 If you have to guess, first narrow the possibilities.

MATH REVIEW — MULTIPLICATION

	2	3	4	5	6	7	8	9	10	11	12
2											
3											
4											
5											
6											
7											
8											
9											
10											
11											
12											

Done in 60 seconds.

6 x 8 = ________

4 x 9 = ________

11 x 12 = ________

9 x 7 = ________

4 x 12 = ________

6 x 3 = ________

8 x 7 = ________

9 x 8 = ________

6 x 7 = ________

6 x 11 = ________

9 x 3 = ________

8 x 12 = ________

7 x 12 = ________

3 x 4 = ________

2 x 9 = ________

7 x 5 = ________

8 x 9 = ________

4 x 5 = ________

6 x 12 = ________

4 x 7 = ________

Cover these answers with scratch paper.

6 x 8 = 48

4 x 9 = 36

11 x 12 = 132

9 x 7 = 63

4 x 12 = 48

6 x 3 = 18

8 x 7 = 56

9 x 8 = 72

6 x 7 = 42

6 x 11 = 66

9 x 3 = 27

8 x 12 = 96

7 x 12 = 84

3 x 4 = 12

2 x 9 = 18

7 x 5 = 35

8 x 9 = 72

4 x 5 = 20

6 x 12 = 72

4 x 7 = 28

MATH REVIEW — WARM UP

Simplify:	Cover these answers with scratch paper.
$\frac{3}{5} \times \frac{5}{4}$	$\frac{1}{2}$
$\frac{.66}{.2}$	3.3
$\frac{1}{3} - \frac{1}{4}$	$\frac{1}{12}$
$2^3 \times 2^7$	2^{10}
$5^3 \times 4^3$	2^3
30% of 250	75
$\sqrt{121}$	11
3×27	9
200×500	100,000
$.4 \times .3$	.12
$\sqrt{50}$	$5\sqrt{2}$
3.21×10^5	321,000

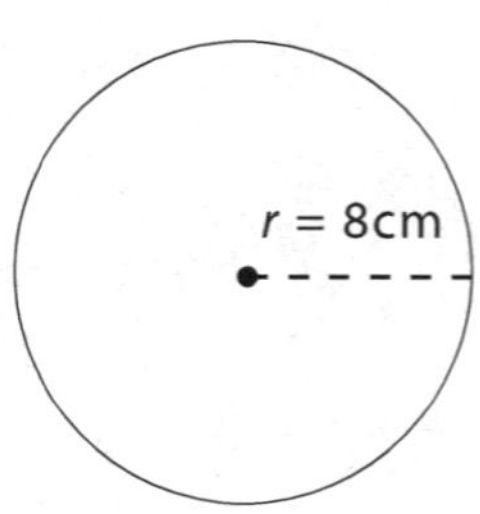

Circumference of this circle ($2\pi r$)	16π
Area of this circle (πr^2)	64π

MATH REVIEW — DRILL | FACTORING TABLE

Directions

1) List all the prime factors
2) List all the factor pairs

For Example:

70   7   5   2

2	35
5	14
7	10

4 ____ ____

6 ____ ____

8 __ __ __
____ ____

9 ____ ____

10 ____ ____

12 __ __ __
____ ____
____ ____

14 ____ ____

15 ____ ____

16 __ __ __ __
____ ____
____ ____

18 __ __ __
____ ____
____ ____

20 __ __ __
____ ____
____ ____

21 ____ ____

22 ____ ____

24 __ __ __ __
____ ____
____ ____
____ ____

25 ____ ____

26 ____ ____

27 __ __ __
____ ____

28 __ __ __
____ ____
____ ____

30 __ __ __
____ ____
____ ____
____ ____

32 __ __ __ __ __
____ ____
____ ____

33 ____ ____

34 ____ ____

35 ____ ____

36 __ __ __ __
____ ____
____ ____
____ ____

38 ____ ____

39 ____ ____

40 __ __ __ __
____ ____
____ ____
____ ____

42 __ __ __
____ ____
____ ____
____ ____

44 __ __ __
____ ____
____ ____

45 __ __ __
____ ____
____ ____

46 ____ ____

48 __ __ __ __ __
____ ____
____ ____
____ ____
____ ____

49 ____ ____

50 __ __ __
____ ____
____ ____

51 ____ ____

52 __ __ __
____ ____
____ ____

54 __ __ __ __
____ ____
____ ____
____ ____

MATH REVIEW — DRILL | FACTORING TABLE

55 ____ ____

56 __ __ __ __
____ ____
____ ____
____ ____

57 ____ ____

58 ____ ____

60 __ __ __ __
____ ____
____ ____
____ ____
____ ____

62 ____ ____

63 __ __ __
____ ____
____ ____

64 __ __ __ __ __ __
____ ____
____ ____
____ ____

65 ____ ____

66 __ __ __
____ ____
____ ____
____ ____

66 __ __ __
____ ____
____ ____

69 ____ ____

70 __ __ __
____ ____
____ ____
____ ____

72 __ __ __ __ __
____ ____
____ ____
____ ____
____ ____
____ ____

74 ____ ____

75 __ __ __
____ ____
____ ____

76 __ __ __
____ ____
____ ____

77 ____ ____

78 __ __ __ __
____ ____
____ ____
____ ____

80 __ __ __ __ __
____ ____
____ ____
____ ____
____ ____

81 __ __ __ __
____ ____
____ ____

82 ____ ____

84 __ __ __ __
____ ____
____ ____
____ ____
____ ____
____ ____

85 ____ ____

86 ____ ____

87 ____ ____

88 __ __ __ __
____ ____
____ ____
____ ____

90 __ __ __ __
____ ____
____ ____
____ ____
____ ____

91 ____ ____

92 __ __ __
____ ____
____ ____

93 ____ ____

94 ____ ____

95 ____ ____

96 __ __ __ __ __ __
____ ____
____ ____
____ ____
____ ____
____ ____

98 __ __ __
____ ____
____ ____

99 __ __ __
____ ____
____ ____

100 __ __ __ __
____ ____
____ ____
____ ____
____ ____

PRACTICE TEST 1

TEST REVIEW — WRONG QUESTION TRACKER

On the practice test—For each question you miss:

1. Why did you miss it?
2. What kind of question was it?
3. Put a check mark if it was Data Sufficiency.

(Don't count click-to-the-end questions.)

Find your gaps and close them BEFORE test day.

	Mistake: Simple math	Mistake: Misread question	Mistake: Clicked wrong answer	Comprehension: Worked it wrong	Comprehension: Didn't know what to do	Comprehension: Didn't know what it's asking	**Question Type** (triangles, etc.)	✓ DS
1								
2								
3								
4								
5								
6								
7								
8								
9								
10								
11								
12								
13								
14								
15								
16								
17								
18								
19								
20								
21								
23								
24								
25								
26								
27								
28								
29								
30								
31								
32								
33								
34								
35								
36								
37								
Total								

PRACTICE TEST 2

TEST REVIEW — WRONG QUESTION TRACKER

On the practice test—For each question you miss:

1. Why did you miss it?
2. What kind of question was it?
3. Put a check mark if it was Data Sufficiency.

(Don't count click-to-the-end questions.)

Find your gaps and close them BEFORE test day.

	Comprehension			**Mistake**				
	Didn't know what it's asking	Didn't know what to do	Worked it wrong	Clicked wrong answer	Misread question	Simple math	**Question Type** (triangles, etc.)	✓ **DS**
1								
2								
3								
4								
5								
6								
7								
8								
9								
10								
11								
12								
13								
14								
15								
16								
17								
18								
19								
20								
21								
23								
24								
25								
26								
27								
28								
29								
30								
31								
32								
33								
34								
35								
36								
37								
Total								

Pretest Guidance

What to do the night before—and the morning of—your test.

Nervous?

Sure you are. So is everyone else—
but you're better prepared than they are.

Remember:

Take your time on the questions.
Rushing kills your score.
Get the first half right, then click to the end.

Essay:

Write your notes before writing the essay.*

Integrated reasoning:

Don't get lost in the data: read the question first.*

Verbal questions:

Don't look for the right answer—eliminate the wrong answer.

Reading Comp and Critical Reasoning:

1) Cover up the answer choices
2) Answer the question yourself
3) Eliminate the wrong answers

* Workbooks coming in 2013

PRE-TEST STRATEGY

The night before your test:

1) Don't study anything: get some rest

You're not going to extract one more point by cramming the night before. Instead, you need your strength. Remember, stamina is a core foundation for success.

2) Eat a good meal, like pasta

Stay away from exotic, unusual foods that might upset your stomach the next day. All you need is that kind of distraction during your test.

3) RELAX: Swim, exercise, or see a movie

Just don't study.

The morning of your test:

1) Do your normal morning routine

— No extra/less coffee
— Good breakfast, but not fancy

I once had two students debate whether to drink more or less coffee on the day of the test. One said he wanted more energy; the other said he didn't want to be nervous. I told them that what they did today will work on test day—in other words, don't change the routine.

2) Warm up

— Simple math/Use the Math Warmup
— Simple reading/Newspaper, web site

Make sure that the first math or verbal question that you get of the day is not a GMAT question. The warmup is effective even if it's hours before. This is like running laps before the marathon.

3) Wear two layers

Some testing centers crank the A/C. If you get cold easily, this can be distracting.

4) Bring food and a bottle of water

You have a 10-minute break, and you don't want the vending-machine food or drinking-fountain water. Bring healthy food, like a nutrition bar or a sandwich, that doesn't need a fridge. If you're like me, you'll also bring a to-go cup of coffee.

Note: You can't have your water with you when taking the test. If this is a distraction, take the practice test without water nearby, so that you become used to it.